Laura Voss
More Than Machines?

Science Studies

Laura Voss is a science manager in the Research Strategy Unit of Ludwig-Maximilians-Universität München. With a background in neuro-cognitive, organizational, and occupational psychology, she previously worked as a science and technology studies researcher and as a science manager at the Munich Center for Technology in Society (MCTS) and for international robotics R&D consortiums at the Technische Universität München.

Laura Voss

More Than Machines?

The Attribution of (In)Animacy to Robot Technology

[transcript]

This work was accepted in 2020 as a doctoral dissertation by the Munich Center of Technology (MCTS) of the Technical University of Munich under the title "Robots Wanted - Dead And/Or Alive. The Attribution of In/Animacy to Robot Technology".

This research was, in part, funded by the Andrea von Braun Foundation.
The Andrea von Braun Foundation aims to contribute to the dismantling of barriers between disciplines, with particular emphasis on the cooperation of fields of knowledge that normally have no or only very little contact with one another. The idea is to create opportunities for the mutual enrichment and cross fertilization of ideas, thus opening access to new and often surprising results and understanding.

Andrea von Braun Stiftung

voneinander wissen

Bibliographic information published by the Deutsche Nationalbibliothek
The Deutsche Nationalbibliothek lists this publication in the Deutsche Nationalbibliografie; detailed bibliographic data are available in the Internet at http://dnb.d-nb.de

Cover layout: Maria Arndt, Bielefeld
Printed by Majuskel Medienproduktion GmbH, Wetzlar
Print-ISBN 978-3-8376-5560-5
PDF-ISBN 978-3-8394-5560-9
https://doi.org/10.14361/9783839455609

Printed on permanent acid-free text paper.

Contents

Acknowledgements

This book is a slightly edited version of my doctoral dissertation, which I defended in June 2020 at the Technical University of Munich (TUM).

I would like to express my heartfelt gratitude – for their guidance, encouragement, and support – to Prof. Dr. Sabine Maasen, Prof. Dr. Jan-Hendrik Passoth, Lisa Meinecke, Stephan Voß, Eva Scheder-Voß, and Prof. Dr. G. Günter Voß.

I am very grateful to the Andrea von Braun Foundation, which kindly supported my research with a scholarship from 2015 to 2017.

Preface

2400 years ago

"The King looked at it in amazement; it was striding quickly looking up and down; undoubtedly it was a man. When the craftsman pushed its cheek it sang in tune; when he clasped its hand it danced in time; it did innumerable tricks, whatever it pleased you to ask. The King thought it really was a man." (Lièzǐ, 列子, ca. 400 B.C.)[1]

Today

"It's a pile of aluminum and copper wire and software. I don't cheer for my laptop. But people cheer for these [robots]. And of course when it falls, we all feel terrible, 'Uh, it got hurt.' But at the end of the day ... It's just a machine." (Gill Pratt, then program manager of the US Defense Advanced Research Projects Agency (DARPA), discussing robots competing at the DARPA Robotics Challenge, 2015)[2]

"I know it's a machine. [But] there was just something about it. It was more reliable than the other ones. ... I just had some connection to it." (US Air Force Colonel Stephen Jones describing a RQ-1 Predator remotely piloted aircraft)[3]

1 Cited in Richey, 2012, p. 194.
2 Cited in Guizzo & Ackerman, 2015.
3 Cited in Pawlyk, 2019.

"*Me, a scientist, not easily fooled. Also me: why are you sad, Cozmo?* 💔" (Anna Henschel, a PhD candidate in Psychology and Human-Robot Interaction, commenting on a Cozmo toy robot in a Twitter post, 2019)[4]

"*We got a Roomba and I get it now ... I would die for this hardworking little man.*" (Ryan Boyd, a computer scientist, commenting on a vacuum cleaning robot in a Twitter post, 2019)[5]

---

4 https://twitter.com/annahenschel/status/1197561427994828800 (accessed 2019-12-02).

5 https://twitter.com/ryandroyd/status/1103782256638812161 (accessed 2019-03-08).

1. Robots Wanted – Dead And/Or Alive

1.1. Making Love and Killing People: The Old and New Age of Robotics

Robots have the connotation of a futuristic technology. In fact, however, they have been around for quite a while: Simple self-operating machines, so-called automata, existed already in ancient Greece, and the manufacturing tradition continued on into medieval times (Truitt, 2015). In the fifteenth century, Leonardo da Vinci drew plans for a humanoid robot (Moran, 2006), and in the eighteenth century Jacques de Vaucanson built his legendary mechanical defecating duck (Riskin, 2003). These automata were mostly toys or pieces of art, bespoke single pieces made not to take on work, but to entertain, to be admired, or to serve as proof for a mechanical concept. Only in the twentieth century had the state of the art, in what was now called robot technology, progressed far enough to be applied on a larger, commercial scale. From the first moving assembly line in a Ford factory in the early twentieth century (Ford.com, n.d.) it was a short way to fully autonomous robots. The earliest concepts for industrial robots emerged in the 1930s. In 1960, the first programmable digital robot was introduced. The 1970s spawned not only the first robotic production lines, but also the first real humanoid robots. Since the early twenty-first century, there even are robots in space and on Mars.

In recent years, robots have been making another important step. They have made their way out of their factory cages and out of robotics laboratories, entering private homes and public spheres to be employed in close physical and social proximity to humans. Today, robotics is a global industry with a 50-billion-dollar turnover. In 2019, 17 million household service robots and 400,000 industrial robots were sold – in addition to the two million already in use (IFR, 2019; Siciliano & Khatib, 2016).

Until recently, the vast majority of robots was employed in the manufacturing industry, and confined to factory cages. They were simply too "dumb" and inflexible, and therefore too dangerous for humans to be around. Since the 2010s, this has begun to change drastically, heralding a "New Age of Robotics" (e.g. Hessman, 2013; Macdonald, 2013):

> "From a largely dominant industrial focus, robotics is rapidly expanding into human environments … Interacting with, assisting, serving, and exploring with humans, the emerging robots will increasingly touch people and their lives." (Siciliano, 2013, p. v)

This new generation of robots is smaller, lighter, more flexible, more adaptive, and more precise – and much more suited for use in close physical proximity to humans, or even in collaboration with them:

> "New robotics no longer concerns only factory applications, but also the use of robotics in a more complex and unstructured outside world, that is, the automation of numerous human activities, such as caring for the sick, driving a car, making love, and killing people." (Royakkers & Est, 2015, p. 549)

Coming in the shape of small mobile platforms, lightweight manipulators ("robot arms") or even with a design inspired by the human body (humanoid robots), a variety of these new robot models are available on the market today, and many more are being developed in academic and commercial robotics labs around the world. Small logistics robots operate in close proximity to humans, for example in Amazon's warehouses[1] (Simon, 2019). Domestic robots like mops, lawn mowers, and vacuum cleaners – such as the popular Roomba[2] – are a huge market success (Tobe, 2014, 2017). Remote controlled mobile platforms with manipulators have become standard equipment in law enforcement and the armed forces, and are routinely deployed in search and rescue operations, explosive ordnance disposal (EOD), and even combat missions (Nosengo, 2019). Collaborative robots ("cobots"), such as Universal Robots'[3] and Kuka's[4] lightweight arms or Rethink Robotics' "Sawyer"[5], are increasingly employed in a range of commercial contexts.

---

1 http://www.amazonrobotics.com (accessed 2019-10-25).

2 http://www.irobot.com/roomba (accessed 2019-10-25).

3 https://www.universal-robots.com/products (accessed 2019-10-25).

4 https://www.kuka.com/en-de/products/robot-systems/industrial-robots/lbr-iiwa (accessed 2019-10-25).

5 https://www.rethinkrobotics.com/sawyer (accessed 2019-10-25).

Dead And/Or Alive?

This new kind of robot and this new form of interaction with robots appear to touch a nerve in the human mind. We have always been fascinated with objects we know to be inanimate but which, for some reason, appear animate to us. For most of human history, these objects were largely restricted to the world of fiction. Robot technology pulls them into the real world, and into our immediate physical and social environment.

Robots have a range of characteristics causing us to associate them with living beings: They are embodied entities in our vicinity, they act autonomously and unpredictably, they sense and react to their environment, they can be mobile and interactive. Crucially, this association is – at least in most cases – not founded in a false belief that robots are actual living beings. It is present in spite of our knowledge that robots are, in fact, inanimate objects. Robots, it appears, can be perceived as both inanimate and animate at the same time.

There is a plethora of both anecdotes and scientific research showing that humans can attribute various lifelike characteristics to robots, and that this influences their attitudes and behavior towards the robots. The field of human-robot interaction studies (HRI) has been producing a vast number of studies trying to explain and quantify the conditions and circumstances of this phenomenon (some of which we will explore in Chapter 2). They usually do so by "measuring" how different characteristics of a human and a robot influence how the human perceives and behaves towards the robot. However, most of this research only explores human reactions to a very specific kind of robot assumed to trigger the strongest attributions of animacy[6]: humanoid robots and robots with an animal-inspired design[7]. This stands in contrast to those robots already in use outside of factories and robotics laboratories today, most of which do not have a humanoid or animal-inspired design.

Most of this HRI research is not conducted in the field, but instead under somewhat artificial laboratory conditions. After all, the few robots that have already made it out "in the field", those that are available for purchase for the average consumer, do not appear to belong to the fascinating group of interactive "new robots". The popular vacuum cleaner robots, for example,

---

6 Chapter 2, Section 2.2, will explain why this book uses the term "animacy" and not, for example "aliveness", "agency" or "intentionality".

7 Sometimes called "zoomorphic robots".

are nothing more than a small disc shaped vehicle driving across the floor, occasionally bumping into a table leg. The simple shape does not keep humans from developing deep social and emotional connections to them, however, not even from attributing animacy or a personality to them (e.g. Sung, Guo, Grinter, & Christensen, 2007; Forlizzi & DiSalvo, 2006; Forlizzi, 2007; Sung et al., 2008; J. Fink, Mubin, Kaplan, & Dillenbourg, 2012). In fact, many users appear to perceive their vacuum cleaner robots as an entity "sit[ting] somewhere between a pet and a home appliance" (Sung et al., 2007, p. 7). There are reports of customers who "t[ook] them on holiday, unwilling to leave them at home alone" (Kahney, 2003).

> They "express[ed] concern when ... told ... to mail in their Roomba and receive a new one in return... they didn't want a new vacuum. ... They wanted 'Rosie' to be ... healed." (Sitrin, 2016)

> "There are people who actually consider them their companion, even though it's just vacuuming their floor ... People get attached to them and think of them as part of their family. It's almost a pet. It makes them feel like they're not alone." (iRobot spokeswoman Nancy Dussault, cited in Kahney, 2003)

These emotional reactions were not intended by the robots' manufacturers. They were, in fact, surprised by their customers' dedication to the little cleaners:

> "When iRobot created Roomba, we didn't want it to be cute; we wanted people to take it seriously, so we gave it more of an industrial look. 25M home robots later, people still personify their Roomba. Over 80% name their robot and many consider it part of the family." (Colin Angle, CEO and founder of iRobot, 2019)

Cute little household helpers are not the only robots with a surprisingly emotional connection to their users. Robots used by bomb squads for explosive ordnance disposal (EOD) are basically small remote controlled tanks with a grasping device on top. And yet they are perceived by their human operators as more than just a tool. These robots are sometimes even considered to be team members, deserving of a funeral when they get "killed" (e.g. Garreau, 2007; P. W. Singer, 2009; J. Carpenter, 2013; Pawlyk, 2019).

> "Sometimes [the soldiers] get a little emotional over it ... Like having a pet dog. It attacks the [bombs], comes back, and attacks again. It becomes part

> of the team, gets a name. They get upset when anything happens to one of the team. They identify with the little robot quickly. They count on it a lot in a mission. The bots even show elements of "personality" ... Every robot has its own little quirks." (Bogosh, cited in Garreau, 2007)

The list of anecdotes goes on: People have been reported to attribute animacy to robotic ottomans (Sirkin et al., 2015), robotic trash cans (Yang et al., 2015), and planetary rovers (e.g. Clancey, 2006; Feltman, 2014; L. Wright, 2016).

That humans can develop emotional connections to technological artifacts is not a new finding per se. It has already been observed with, for example, cars (Chandler & Schwarz, 2010) or mobile phones (Jane Vincent, 2005). The case of robots, however, is unique in that it sparks a new discussion about the ontological status of technological artifacts. In all of the examples explored above, people are very well aware that robots are inanimate objects. Nonetheless, something about robots makes them appear to be more than "just machines".

The question of what robots are, ontologically, has sparked a lively discussion across public and academic discourses. Scholars across disciplines have described robots as "neither alive nor not alive" (Severson & Carlson, 2010, p. 1101), "neither and both" (Melson et al., 2009, p. 563), "alive in some respects and not alive in other respects" (Kahn et al., 2004, p. 549), "both animate and inanimate" (De Graaf, 2016, p. 592), "sort of alive" (Turkle et al., 2004, p. 4), "stand[ing] between an 'animal kind of alive' and a 'human kind of alive'" (ibid., p. 11), or "simultaneously enacted as an agent and as a thing" (Alač, 2016, p. 526). Other authors described robot users as showing "a 'weird' doubleminded attitude" (Bruckenberger et al., 2013, p. 305) or as holding "parallel conceptions" of robots (Fussell et al., 2008, p. 151).

It appears to be quite difficult to sort robots in a dichotomy of "animate" and "inanimate". Some scholars even propose to create an altogether new, different ontological category for them:

> "If from the person's experience of the subject-object interaction, the object is alive in some respects and not alive in other respects, [it] is experienced not simply as a combination of such qualities ... but as a novel entity." (Kahn et al., 2004, p. 549; cf. Severson & Carlson, 2010)

1.2. Hype, Hope, and Horror

While the scholarly discussion on the ontological status of robots is going on in academic journals and conference halls, the machines in question are already making their mark on our everyday lives. Almost every day the news report on yet another revolutionary robot technology being "unleashed" on society. It seems that, finally, all the robots we so far only knew from science fiction are becoming reality – as are the scenarios associated with them, both the hopeful and the scary ones.

The current hype around robot technology is thoroughly embedded in – and fueled by – culturally shared visions and imaginaries of a robot-populated future. Crucially, these visions cannot only be encountered in popular culture, in science fiction movies, shows, and novels. They are very much a part of our "real life", in that they shape the way robot technology is discussed by laypeople, by the media, and by policy makers.

The notion of visions of the future crucially shaping the development of emerging technologies has been of interest for science and technology studies (STS), sociology, and innovation studies. Concepts like guiding visions (German "Leitbilder"; cf. e.g. Giesel, 2007), expectations (e.g. Borup et al., 2006; Beckert, 2016; Brown et al. 2000), socio-technical imaginaries (e.g. Jasanoff & Kim, 2009, 2015), or socio-technical futures (e.g. Böhle & Bopp, 2013) describe how imaginations of the future shape the development of technology in the present. These different approaches all share the idea that visions of the future have a guiding and structuring function. By drawing the focus on a shared horizon and "[preparing] possibilities of future events" (Luhmann, 1988, p. 121), they reduce the complexity of possible paths into the future. They help to define roles and tasks, and to legitimize and coordinate science and governance efforts, such as resource allocation and legislation (Borup et al., 2006; Giesel, 2007).

These functions can also be observed for the case of robot technology. Visions for a robotized society can be found in policy documents across the world. In the European context, there is the EUROP[8] Strategic Research Agenda, with elaborate "Product Visions & Application Scenarios" (EUROP, 2006, 2009), or the SPARC[9] Strategic Research Agenda, featuring short stories on desirable robot applications (SPARC, 2013). There is Japan's New Robot

8 European Robotics Platform (http://www.robotics-platform.eu).

9 "The Partnership for Robotics in Europe" (https://www.eu-robotics.net/sparc).

Strategy (The Headquarters for Japan's Economic Revitalization, 2015), the Korean Robot Act (cf. S. Kim, 2018), and the US National Robotics Roadmap (Computing Community Consortium, 2016a). They all lay out visions for futures in which robot technology is employed in almost all possible contexts – from health care to education, from manufacturing to transportation.

While these documents are mostly geared towards policy makers, other – more extreme – imaginations draw much public attention. With robotics being "an emerging and significant area of controversial technoscientific development" (Wilkinson, Bultitude, & Dawson, 2011, p. 373), public discourse appears to be split between two contrasting narrative poles. At one end of the spectrum, we can observe a discourse steeped with utopian techno-optimism, promising solutions to a score of societal problems. At the other end, there is a dystopian-pessimistic discourse, dominated by a view of robots as competition for humanity, by fears of humanity being replaced or subjugated by ultra-intelligent and powerful "robot overlords". Both of these discursive extremes are heavily influenced by popular science fiction tropes (cf. COMEST, 2017, p. 40; Gesellschaft für Informatik, 2019), which we will explore in more detail in Section 1.3.

The impression left by this discourse, that robot technology is getting "closer to science fiction", is in fact not completely unfounded. Most of the robot characters we know from popular science fiction stories are able to act autonomously, interact socially with humans, have roles and tasks traditionally reserved for humans, and often even look like humans. This – as discussed above – is exactly the kind of "New Robots" which have been making their way from robotics laboratories into our everyday lives.

It is also the kind of robot technology that has been showered with political and financial support in the recent past. In the United States, the 2011 National Robotics Initiative (NRI) and its 2016 successor NRI 2.0 dedicated around 100 million dollars of funding "to accelerate the development and use of robots in the United States that work beside or cooperatively with people" (Jahanian, 2011; National Science Foundation, 2019). Japan's 2015 New Robot Strategy and Robot Revolution Initiative, with service robotics as one of the funding foci, pushed to quadruple the Japanese robotics market to 2.4 billion yen by 2020 (The Headquarters for Japan's Economic Revitalization, 2015; Edwards, 2015). The European Commission poured 700 million Euros into SPARC, "the largest civilian robotics research and innovation programme in the world" (EU Robotics, 2018). This funding boost is part of a global robotics "arms race", fueled by a political and economic discourse that constructs robot technology as inherently useful (Bischof, 2017a, p. 138). As Andreas Bischof notes, research and development in these initiatives is often driven by a goal of finding problems to which to apply robot technology – rather than the goal of finding robot-assisted solutions to existing practical problems (2015, pp. 156 & 181). The idea of an inevitable future brimming with robot technology seems to serve as a self-fulfilling prophecy (cf. Meister, 2011, p. 120).

The underlying assumption of usefulness, hope, and even salvation through robots is also observable at the "utopian extreme" of public discourse. Here, robots are presented as a universal solution for some of today's most pressing issues. Most prominently among those issues: the "alarmist demography" alerting to an aging society in an overpopulated world (Katz, 1992), but also environmental and health crises, unemployment, armed conflict, and more. Robot technology is praised as a "new solution ... to societal challenges from aging to health, smart transport, security, energy and environment" (European Commission, 2015), with "the potential to transform lives and work practices, raise efficiency and safety levels, provide enhanced levels of service and create jobs" (SPARC, 2013, p. 6). In this, robots are understood to "represent the dawn of a new era, ubiquitous helpers improving competitiveness and our quality of life" (SPARC, 2013, p. 15). Robot technology is almost hailed as a panacea:

> "Robots can save lives and reduce the economic consequences of disasters ... Home health care, mobility, wellness and well-being are being positively impacted by assistive robotics, human-robot interaction, advanced prosthetics, and smart sensing ... Robotics can be seen as a tool for not just enhancing but potentially revolutionizing K-12 STEM education ... low[ering] the digital divide, and bring[ing] more gender and ethnic balance to the STEM workforce. ... Social robots can boost the confidence and self-esteem of children from all socio-economic backgrounds." (Report on US robotics development during the National Robotics Initiative, Computing Community Consortium, 2016b, p. 4)

Across the whole spectrum of discourse, from the pessimistic-dystopian to the hopeful-utopian visions, the increasing application of robot technology in an unspecified future is presented as self-evident and unquestionable. As Andreas Bischof notes, policy documents and public discourse are saturated with a "fatalistic conviction of the unavoidability of a robotic future"[10] (2017a, p. 163), a "teleological inevitability and desirability, inferred from a desire for technical feasibility"[10] (2019; cf. Jasanoff & Kim, 2009). The European Commission and the European Robotics Platform EUROP describe robotized futures as if they were already set in stone, predicting that "as assistants, robots will be co-workers in the workplace, companions at home, servants, playmates, delivering professional services and acting as agents for security" (European Commission, 2008, p. 4), and that "in the service sector robotics coworkers will assist humans performing services useful to the well-being of humans or equipment" (EUROP, 2009, p. 15). Technical and social challenges and obstacles are downplayed, met with counter-arguments, or simply negated. For example, the prominently and controversially discussed issue of technological unemployment – increasing automation potentially making a human workforce obsolete – is quickly settled in a policy document by the European Commission: "While the installation of robots may result in immediate redundancies, the long-term benefits to employment cannot be denied" (European Commission, 2008, p. 1). Other challenges, too, are often swept under the carpet in these and similarly enthusiastic publications: technical bottlenecks like battery capacity, natural language interfaces, or unstructured environments, but also non-technical issues like user acceptance or ethical concerns.

The predominantly utopian policy discourse stands in stark contrast to a much more dystopian, albeit often similarly fatalistic, parallel discourse. Especially in the news media, enthusiastic reports on new robotic technologies are neighbored by predictions of widespread unemployment, of humans being replaced, even of robots "going rogue" and rising as "robot overlords". Also political discourse is peppered with references to this dystopian narrative, especially in the context of autonomous weapons ("killer robots"; e.g. Human Rights Watch, 2014; Sychev, 2018).

This type of controversial and emotional rhetoric can be observed for other emerging technologies as well, which likewise "exist in a state of flux as a mixture of blueprint and hardware, plan and practice, … surrounded by speculation and speculators, who make often-contested claims about their promises, perils, and possibilities" (Hilgartner & Lewenstein, 2014, p. 1). The neighboring discourses on artificial intelligence, machine learning, and neural networks are just as torn between hopes and fears (Marcus, 2013; Cave & Dihal, 2019). The discourse on genetic technology is similarly infused with a "discourse

10 Translated from German by the author

of great promise [and] great concern" (Tambor et al. 2002, p. 35; cf. Durant, Hansen, & Bauer, 1996).

Among the many emerging and controversially discussed technologies robotics has a prominent position, however, as it often is perceived as standing paradigmatically for technological progress. It seems as if robot technology is somehow "always in the future. Every once in a while, a piece of it breaks off and becomes part of the present" (Huggins, cited in Loukides, 2013). In this, robotics is an example of how a strong existing vision of a technological future can also hinder developmental flexibility (cf. Dierkes, 1988, p. 58): "Once technical promises are shared, they demand action, and it appears necessary for technologists to develop them, and for others to support them" (Van Lente & Rip, 1998, p. 17). This makes it possible to misuse robots as a "technofuturistic escape" (Jeon, 2016): Promoting idealized scenarios of a future in which today's pressing problems have been solved by robot technology conveniently gives policy-makers the possibility to evade addressing the current problems.

1.3. Robots and Science Fiction: Inseparably Linked

Robotics' curious status as a technology that is both futuristic and well known to everyone and its prominent discursive position as an unavoidable technology are rooted in the fields' unique history. In robotics, fictional narratives and real technological progress have always advanced hand-in-hand: "No technology has ever been so widely described and explored before its commercial introduction" (Jordan, 2016, p. 5). While technological predecessors of today's robots can be traced back to medieval automata (Truitt, 2015; cf. Section 1.1), the idea of autonomously acting "animated" objects goes back even further, to Greek, Byzantine, and Chinese myths (Brett, 1954; Needham, 1956; "Automatones," n.d.).

This historically close connection of fictional narratives and technological development can be observed in robotics until today. Not only does the word "robot" stem from a 1920 theater play (Čapek, 1920), science fiction narratives also crucially influence roboticists' identity and everyday life culture (Bischof, 2017a, p. 141). They serve as what Susan Leigh Star and James Griesemer (1989) called a "boundary object". Both within the robotics community and in communication efforts by robotics with the general public they provide a shared discursive framework and focus of attention, and they act as a repository for the epistemology of the public discourse on robotics[11]. Even the official website of the IEEE[12], the world's largest association of technical professionals, states that "for most ... [roboticists], science fiction has strongly influenced what [they] expect a robot to look like and be able to do" (Guizzo, n.d.). This is also reflected in the many forewords of robotics handbooks written by science fiction authors – such as Isaac Asimov's forewords for Joseph Engelberger's (1980) "Robotics in Practice" and Shimon Nof's (1985) "Handbook of Industrial Robotics"; in the mission statements of commercially successful robotics companies – such as iRobot, which considered the mottos "making science fiction reality" and "practical science fiction" (P. W. Singer, 2009, p. 185); and in countless implicit and explicit references to science fiction as a "hidden curriculum" for robotics (Bischof, 2017a, p. 145; cf. Rammert, 2001, p. 22).

The connection of robotics and science fiction goes so far that policy actors explicitly base legislative decisions on science fiction narratives (Chapter 6 will explore this issue in depth). The most prominent – and controversial – examples are probably the many references to Asimov's (1950) "Three Laws of Robotics"[13] in the discourse on robotics and artificial intelligence (AI) legislation. Even human rights experts, in interviews on the laws of unmanned warfare, have been noted to "reference ... Blade Runner[14], the Terminator[15], and Robocop[16] with the same weight as ... the Geneva Conventions" (P. W. Singer, 2009, p. 203).

The common practice of understanding science fiction as inspiration, even as a blueprint, for real-life technology development, governance, and legislation faces considerable criticism. Several science fiction writers felt the need to emphasize that their stories are not to be understood as predictions or even recommendations. For example, award-winning science fiction author Ursula K. Le Guin (1976) cautioned:

11 Hat tip to Lisa Meinecke.

12 Institute of Electrical and Electronics Engineers (https://www.ieee.org).

13 The Three Laws of Robotics: "(1) A robot may not injure a human being or, through inaction, allow a human being to come to harm. (2) A robot must obey the orders given it by human beings except where such orders would conflict with the First Law. (3) A robot must protect its own existence as long as such protection does not conflict with the First or Second Laws and (0) A robot may not harm humanity, or, by inaction, allow humanity to come to harm" (Asimov, 1950).

14 The film "Blade Runner" (R. Scott, 1982) is based on a short story by Philip K. Dick (1968).

15 The Terminator is the cyborg protagonist of a successful film franchise, starting with "The Terminator" (Cameron, 1984) and comprising six films as of 2019.

16 Robocop is the cyborg protagonist of the film of the same name (Verhoeven, 1987).

> "Science fiction is not predictive; it is descriptive ... But our society, being troubled and bewildered, seeking guidance, sometimes puts an entirely mistaken trust in [science fiction authors], using them as prophets and futurologists."

It has been argued that, in fact, science fiction is more concerned with the present than the future. Science fiction author William Gibson pointed out:[17]

> "It's about the present. It's not really about an imagined future. It's a way of trying to come to terms with the awe and terror inspired ... by the world in which we live." (cited in Leary, 1989, p. 58)

Understanding science fiction as inspiration for the development of new technologies disregards that most science fiction stories are a way of reflecting on, even criticizing, the past and present reality. As Lisa Meinecke and I noted:

> "[Science fiction] is not a neutral repository of ideas about technology or a roadmap to the future. The narratives are shaped by the cultural context they originate from, by the values, hopes, and anxieties of society. ... A fictional robot is rarely just a robot, it is also a narrative canvas for projections of the other, which carries a culture's hopes and anxieties." (Meinecke & Voss, 2018, p. 208)

This reflexive and critical aspect of science fiction appears to go over the heads of many scientists, innovators, and policy makers who seem to understand science fiction as mostly inspirational. A 2018 cartoon commented on science-fiction-inspired expectations of future robots by contrasting an "unlikely" scene from the Terminator movie (Cameron, 1984) with a "likely" alternative concept, in which not the socially interactive cyborg Terminator but a small non-humanoid wheeled platform seeks out the other main protagonist Sarah Connor (see Figure 1).

William Gibson, whose novels often feature an extremely dark and violent technological future, experienced this as well, with "analysts and politicians ... actively draw[ing] on [his novels] to justify investment in information and communications technologies" (Kitchin & Kneale, 2001, p. 24). He noted: "The social and political naivete of modern corporate boffins is frightening, they read me and just take bits, all the cute technology, and miss about fifteen levels of irony" (cited ibid.).

---

17 Specifically referring to his cyberpunk novel "Neuromancer" (1984).

Figure 1: Cartoon "Robot Apocalypse" (XKCD, 2018).

Source: https://what-if.xkcd.com/5 (accessed on 2019-11-26). Image used in accordance with the artist's guidelines (https://xkcd.com/license.html).

In fact, there is "a recursive relationship between scientific every day practice and fictional technology futures" (Bischof, 2017a, p. 145). Not only do fictional narratives influence real-life robotics, many newer science fiction stories reference current developments in robot technology (Meinecke & Voss, 2018, p. 206). This "sci-fi feedback loop" is not unique to robotics but can also be observed, for example, with space flight and defense technologies (Bankston & Finn, 2019).

The process of "science unfiction" (Poon, 2000) goes so far that technology companies and even government agencies hire science fiction writers as consultants. In 2019, the French army announced the creation of a team of science fiction writers, whose task it would be to "propose scenarios of disruption that military strategists may not think of" (BBC News, 2019). Moreover, science fiction films have been shown to feature "diegetic prototypes" of, for example, space ships – providing scientific organizations with promotional images and arguments for the necessity and viability of new space flight technology (Kirby, 2010).

Science fiction narratives not only serve as a boundary object for roboticists themselves, but also for their communication with the lay public (Šabanović, 2007). Due to its immense popularity, science fiction strongly influences most people's ideas of what robots look like and what their capabilities are (Gesellschaft für Informatik, 2019; Bruckenberger et al., 2013). The emotional and controversial public discourse on robotics, as discussed above, reflects this science fiction-fed notion of robots. The discourse is torn between utopian and dystopian conceptions of a robot-populated future, and constantly refers to the recurring themes of robot science fiction: the robot as an "other", and the question what constitutes a human being; robots, androids, and cyborgs as more or less elaborate artificial humans; robots being treated (or not) like humans; robots wanting to become humans; robots standing in competition to humans; robots wanting to overthrow humanity – all these are staples of past and present science fiction literature, TV, and cinema (Meinecke & Voss, 2018). We find these themes in early android stories like Fritz Lang's (1926) "Metropolis", Isaac Asimov's (1950) "Runaround", or Philip K. Dick's (1968) "Do Androids Dream of Electric Sheep?"; in 1980s Hollywood movies like "Blade Runner" (R. Scott, 1982; an adaptation of "Do Androids…"), "The Terminator" (Cameron, 1984), "Robocop" (Verhoeven, 1987), or "Short Circuit" (Badham, 1986); and in movies and TV series of the 2010s, like "Ex Machina" (Garland, 2014), "Big Hero Six" (C. Williams & Hall, 2014), "Chappie" (Blomkamp, 2015), or "Westworld" (Nolan & Joy, 2016). All these stories are expressions of humans' long-standing fascination with stories of "objectively" inanimate machines with characteristics of living beings. With recent developments in the "New Age of Robotics", this fascination is not restricted to the world of fiction anymore (cf. Section 1.1). There are now real robots coming staggeringly close to what, so far, was only known from fiction:

> "Robotics as a technology is fascinating because it represents, even just in the last 20 years, this transition of an idea from something that's always been [relegated to] pop culture to something that's real." (Wilson, cited in LaFrance, 2016)

1.4. Research Question and Approach

The previous sections showed that robot technology – especially the kind of "New Robotics" that is increasingly interactive and employed in close physical and social proximity to humans – is at the center of a controversial discourse spanning fictional narratives, academic research, political decision-making, and public discourse. Robots' polarizing position as either our companions, coworkers, even saviors on the one hand, and as our competition, even potential oppressors on the other hand, is closely tied to their disputed ontological status. Are they "only" inanimate machines? Or are they – somehow – more similar to living beings than other technological artifacts?

This ontological problem, which is so heavily reflected across discourses, also fuels the overarching question (or rather, questions) this book wants to explore. Crucially, this book does not aim to decide what robots "really are", whether robots are like living beings or not, or whether robots should be made to resemble living beings or not. Rather, it will follow an idea voiced by Lucy Suchman, who in her book "Human-Machine Configurations" proposed to

> "[shift the discussion from] whether humans and machines are the same or different to how and when the categories of human or machine become relevant, how relations of sameness or difference between them are enacted on particular occasions, and with what discursive and material consequences." (Suchman, 2007, p. 2)

Inspired by this proposal, we will explore two overarching questions:

1. **Which discursive and non-discursive manifestations of in/animacy attributions to robots are there?**
2. **What are the conditions, functions, and consequences of these attributions?**

In the spirit of Foucault's (1977) dispositive analysis, we will take apart the complex apparatus of "discourses, institutions, ... regulatory decisions, laws, ... scientific statements, philosophical, moral and philanthropic propositions" (ibid., p. 194), exploring "the said as much as the unsaid" (ibid.).

This approach is meant to set an explicit contrast to existing research that only looks at a very narrow aspect of robotics – usually that of physical human-robot interaction. We will instead explore the whole "life cycle" of robots, ranging from their conception in fictional and nonfictional visions of a robotized future, to practices of making them physical reality in research and development, to their presentation to different audiences in demonstrations, science communication, and marketing, to users' actual interaction with robots, to the reception of robot technology in the medial and political discourse (see Figure 2).

By following this cycle, we can utilize robot technology as an entry point for the exploration of our present technologized society, in which new technologies fundamentally permeate, and constantly challenge, the lives of individuals, collectives, and organizations. Technologies such as mobile

Figure 2: The "life cycle" of robots.

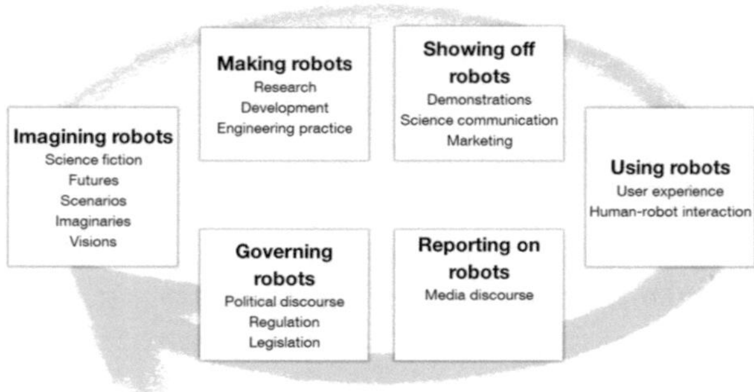

communication, neurotechnology, or advanced prosthetics, which, in the words of Donna Haraway, "have made thoroughly ambiguous the difference between natural and artificial, mind and body, self-developing and externally designed, and many other distinctions that used to apply to organisms and machines" (1991, p. 152). Robotics is also one of these technologies. Robots can be intelligent, embodied, and autonomous – all characteristics we traditionally know from living beings. They thus challenge traditional views of who and what can be a communicative and socially interactive "other", and appear to scrape, or even break, the boundaries of the social world (cf. Luckmann, 1970; cf. Lindemann, 2005).

1.5. Some Methodological Clarifications

Interdisciplinarity

While guided by the conceptual and methodological traditions of science and technology studies (STS), this book was written with an interdisciplinary mindset. It is aimed at readers of all disciplines and cites relevant literature from a wide range of academic fields. This is not only because the author does not consider herself belonging to one specific discipline: a neuro-cognitive psychologist by training, advised by a sociology of science professor, with professional experience as a researcher and as a science manager, gained at an interdisciplinary research center for science and technology studies and in robotics research and development. It is also also because the issue of what (or who?) robots are – ontologically, culturally, perceptually – is of interest for and researched by a wide range of disciplines, including, but not limited to, science and technology studies, cognitive and evolutionary psychology, communication studies, anthropology, philosophy, and human-machine interaction studies.

Definition of "Robot"

The question of "what a robot is" is difficult to answer – and not only because of robots' unclear ontological status. Even the IEEE provides no official definition of "robot". Its website notes that "the term 'robot' means different things to different people. Even roboticists themselves have different notions about what is or isn't a robot" (Guizzo, n.d.). This issue has become somewhat of a running joke, as the following quotes illustrate:

> "I asked some very smart people a pretty simple question, at least on the surface: 'What is a robot?'. I received answers dripping in ambiguity." (Pearson, 2015)
>
> "Never ask a roboticist what a robot is. The answer changes too quickly. By the time researchers finish their most recent debate on what is and what isn't a robot, the frontier moves on as whole new interaction technologies are born." (Nourbakhsh, 2013, p. xiv)
>
> "Ask three different roboticists to define a robot and you'll get three different answers." (Simon, 2017a)
>
> "I don't know how to define [robot], but I know one when I see one!" (Robotics pioneer Joseph Engelberger, cited in Guizzo, n.d.)

Within the robotics community, opinions differ on whether a robot, in order to be called a robot, has to be mobile (which would exclude stationary robot arms), autonomous (which would exclude remote controlled robots), or interactive (which would exclude many industrial robots). Even the most basic consensus – describing a robot as a machine that can sense, compute, and act in the physical world – is problematic, as this would include seemingly "unrobotic" devices, such as dishwashers and thermostats (Guizzo, n.d.).

In public discourse, even the idea of embodiment is debatable (as Chapter 5 will discuss in depth). On the one hand, many laypeople only think of humanoid robots when they hear "robot", as this is what they know from science fiction. On the other hand, the word "robot" is routinely used in media discourse for non-embodied ("virtual") technologies such as software, AI, and chat bots, or even automation in general (cf. LaFrance, 2016).

For the purpose of this book, the question of which machines are considered robots will be approached like the question of what ontological category robots belong to: Following Lucy Suchman's (2007, p. 2) idea, we will not dwell on which machines "really" are robots, but instead explore when and where the concept of a robot becomes relevant, how it is enacted on particular occasions, and with what discursive and material consequences.

A working definition proposed by Neil Richards and William Smart (2013) closely matches this sentiment:

> "A robot is a constructed system that displays both physical and mental agency, but is not alive in the biological sense. That is to say, a robot is something manufactured that moves about the world, seems to make rational decisions about what to do, and is a machine. It is important to note that the ascription of agency is subjective: the system must only appear to have agency to an external observer to meet our criteria. In addition, our definition excludes wholly software-based artificial intelligences that exert no agency in the physical world. Our definition intentionally leaves open the mechanism that causes the apparent agency. The system can be controlled by clever computer software, or teleoperated by a remote human operator." (Richards & Smart, 2013, p. 5)

On a practical level this means: when a field actor calls something a robot, we will consider it to be a robot.

Cultural Context

Not only is robot technology defined differently in different contexts. The discourse on robotics, their use, and their acceptance in society is also influenced by a vast range of variables. One of these variables is the cultural context. Especially a presumed east-west divide of cultural acceptance of robots has received quite a bit of academic attention. The question of whether eastern cultures are more accepting of robots than western cultures, and whether this has to do with a shintoistic understanding of all objects possessing a spirit[18], is at the focus of interest of a plethora of studies[19]. This book will not explore these cultural differences – as relevant and interesting they are – but explicitly stay on the western side of the presumed divide. The investigated cases are all from the European and North American cultural context. This also means that the insights of this book are not necessarily generalizable to other cultural contexts.

1.6. A Tour Along the Life Cycle of Robots

In the next five chapters, we will go on a journey along the life cycle of robots, exploring a range of different contexts in which robots play a role. In the introductory sections of Chapter 1 we already explored how recent technological developments bring robot technology into closer interaction with humans, and how this raises the question of which ontological category robots belong to, of whether robots can be "animate". We also took a step back to what could be understood as the starting point of a robot's life cycle and explored the fictional and real-life visions that have been crucially influencing technical progress in, as well as public and political discourse on, robotics.

Up next, Chapter 2 will equip us for the further progress of our tour by providing some conceptual tools and disciplinary background. We will untangle the complex terminological, conceptual, and historical context of research on attributions of in/animacy to inanimate objects.

Chapter 3 will continue the tour by diving into the representation of robots as in/animate in the context of robotics research and development.

Chapter 4 will explore how robots are presented as in/animate to different expert and non-expert audiences in the context of robotics demonstrations, science communication, and marketing.

Chapter 5 will examine how robot technology is represented as in/animate in the news media.

---

18 In contrast to the Judeo-Christian understanding that only God can give life and animated objects are therefore sinful.

19 E.g. Kaplan, 2004; Bartneck, Nomura, Kanda, Suzuki, & Kato, 2005; Geraci, 2006; Kitano, 2007; MacDorman, Vasudevan, & Ho, 2008; Weng, Chen, & Sun, 2009; Tatsuya Nomura, Sugimoto, Syrdal, & Dautenhahn, 2012; Wagner, 2013; Šabanović, 2014; Kamide & Arai, 2017.

The final Chapter 6 will discuss, now from a cross-contextual perspective, the constructive contributions of in/animacy attributions and the critical discourse we will observe across contexts, and provide some take-home messages for the readers of this book.

2. Disciplinary Context and Terminology

Before we continue our tour along the life cycle of robots and explore how in/animacy is attributed to robot technology in different contexts, we first need to equip ourselves with some conceptual tools.

The question of when and why humans attribute characteristics of living beings to non-living entities, or characteristics of humans to non-human entities, has been a topic of interest for several scientific fields. Within and across these different disciplines a range of terminology is employed to describe the same or similar phenomena. The present chapter will untangle this complex disciplinary, historical, and terminological context.

First, the chapter will show how human-robot interaction (HRI) research approaches the phenomenon of animacy attribution to robots. We will explore the field's strongly innovation- and application-driven approach towards the phenomenon, and explore basic assumptions underlying this research, as well as methodological and ethical issues discussed in this context.

Second, we will take apart the tangle of different terms used across disciplines – such as "anthropomorphism", "animacy", "intentionality", and "agency" – and establish the use of the term "attribution of animacy" for the purpose of this book.

Third, the chapter will give an overview of further relevant disciplinary perspectives on the topic. It will show that, historically, phenomena like anthropomorphism have often been viewed either as a "primitive" interpretation of environmental cues or, in the context of academia, as methodological malpractice. Only relatively recently has the topic drawn scientific interest as an object of research in itself. We will see how different disciplines approach issues like anthropomorphism, animacy detection, and technological agency.

2.1. Human-Robot-Interaction Research: "Controlling" In/Animacy Attributions

In Chapter 1 (Section 1.1), we already touched upon the fact that most academic attention on the phenomenon of animacy attributions to robots can be observed in the context of human-robot interaction (HRI) research. Most HRI research takes place in the field of social robotics, focusing on robots explicitly meant for more complex user interactions, usually with a humanoid design. These robots can be either bespoke platforms or off-the-shelf models like Softbank's Nao[1], which are still too expensive for the average customer. Robot technology safe and robust enough to be employed in direct physical contact with humans is only just now becoming available and affordable enough for the mass market, for example in the form of lightweight robot arms, household robots, or small tele-operated platforms. These robots are usually not intended for complex social interactions. Nonetheless, there is a slowly growing awareness of the complexity of interaction with "mechanical looking" and seemingly "non-social" robots not specifically designed to interact with humans or to appear life-like in any way. Andrea Guzman (2016), for example, argues for the designation of industrial and manufacturing machines as technologies of communication.

The goal of research efforts in HRI is usually not the short-term realization of an interaction scenario representing the current state of robot technology. Instead, HRI research usually focuses on the exploration of scenarios expected to become relevant only in the future, such as the coexistence with very human-like robots (Bischof, 2015, p. 211). In order to simulate the anticipated capabilities of future robots some interaction studies make us of so-called "Wizard-of-Oz" experiments, in which the robot's behavior is secretly controlled by a human operator. This method is sometimes criticized – both for its deception of study participants and for not really studying human interaction with the robot, but rather with the human robot operator, only relayed through a robot (cf. Laurel D. Riek, 2012).

For most people, the long-term use of robot technology, especially socially interactive robot technology, is not yet an everyday practice. Most HRI research therefore studies short-term interactions in laboratory settings, or explores attitudes towards robots based on people's existing knowledge and imagination of robots.

---

1 https://www.softbankrobotics.com/emea/en/nao (accessed 2019-12-21).

Notable exceptions to this approach are studies exploring some of the few contexts where users already closely interact with robots every day. This includes the professional use of remote controlled robot platforms for explosive ordnance disposal (J. Carpenter, 2013, 2016) or search and rescue efforts (Bethel & Murphy, 2006; Murphy, Riddle, & Rasmussen, 2004), and the use of vacuum cleaner robots in private households (Forlizzi & DiSalvo, 2006; Sung et al., 2007). These field studies, which all focus on non-humanoid robots, only superficially deal with the question of how or why humans attribute characteristics of living beings to robots. In the more short-term, laboratory-based research, however, this topic receives plenty of scholarly attention. Especially in the field of social robotics, a range of studies investigates what is sometimes called "anthropomorphic projections" or "anthropomorphic attributions"[2] to robots (Damiano & Dumouchel, 2018, p. 2) – the perception of robots as having human-like characteristics. This research is frequently based on the findings of cognitive and evolutionary psychology – an aspect explored further in Section 2.3 of this chapter. The methodological spectrum and quality of these studies is broad. It ranges from psychophysiological and neuroscience methods, to behavioral observation, to cognitive tests and self-assessment questionnaires.[3] While the majority of these studies uses "homemade" methodological tools, there are some efforts to standardize the "measurement" of anthropomorphic projections, such as the Godspeed questionnaire (Bartneck et al., 2009; Ho & MacDorman, 2010) or the RoSAS scale (Carpinella et al., 2017).

Underlying this approach is an idea inherent to the innovation-driven interests and methods of HRI research: that of "controlling" humans' anthropomorphic attributions to technology. In contrast to the historical skepticism towards anthropomorphism and similar phenomena in other academic disciplines (which we will explore in Section 2.3), the self-imposed challenge for HRI research is "not how to avoid anthropomorphism, but rather how to embrace it" (Duffy, 2003a, p. 180). The goal is to identify user and robot characteristics involved in the "activation" of anthropomorphic attributions. These characteristics are then supposed to act as predictors for specific behavioral

2 Section 2.2 will explain why this book uses a different term for the same phenomenon ("animacy attribution").

3 For a general overview see e.g. Złotowski et al. (2018) or Damiano and Dumouchel (2018). For an overview of neuroscience approaches to human–robot interaction research see Henschel, Hortensius & Cross (2020).

and emotional reactions of the user, which in turn are understood as indicators for the "strength" of the anthropomorphic attribution.

Typical user characteristics used as independent variables in this type of HRI research are standard demographic variables such as age, gender, or cultural background, as well as personality traits.[4] The catalog of variables also includes complex (and difficult to operationalize) traits such as "loneliness", "need for control", "experience with robots" and "interest in technology".[5] On the side of the robot, HRI research explores relatively simple variables like size, color, or material, but also more complex, usually unstandardized factors such as "physical presence", "human likeness", "animacy", or "behavioral complexity".[6]

Combinations of these independent variables are then explored in their effect on various emotional and behavioral measures. Some of these are meant to quantify the "amount" of human-likeness study participants attribute to robots. Here, we find studies observing "intelligence attribution", "mind perception", "perceived social presence", "perceived sociability", but also "embarrassment from being observed by the robot", "empathy with the robot" and "hesitation to switch off", or refusal to physically "harm", or "kill" a robot.[7]

---

4 Selected examples: Age (e.g. Kuo et al., 2009; Reich & Eyssel, 2013); gender (e.g. Chin, Sims, Clark, & Lopez, 2004; De Graaf & Ben Allouch, 2013; T. Nomura, Kanda, Suzuki, & Kato, 2008; Schermerhorn, Scheutz, & Crowell, 2008); cultural background (e.g. Evers, Maldonado, Brodecki, & Hinds, 2008; Tatsuya Nomura et al., 2008); personality traits (e.g. Syrdal, Dautenhahn, Woods, Walters, & Kheng Lee Koay, 2006; Walters et al., 2005; Woods et al., 2007).

5 Selected examples: Loneliness (e.g. Epley, Akalis, Waytz, & Cacioppo, 2008; Reich & Eyssel, 2013); need for control (e.g. Epley, Waytz, Akalis, & Cacioppo, 2008); interest in/experience with robots or technology (e.g. Bartneck, Suzuki, Kanda, & Nomura, 2007; European Commission, 2012; Heerink, 2011, 2011; Tatsuya Nomura, Suzuki, Kanda, Yamada, & Kato, 2011; Reich & Eyssel, 2013; Woods et al., 2007).

6 Selected examples: Size (e.g. Walters, Koay, Syrdal, Campbell, & Dautenhahn, 2013); color/material (e.g. J. Wright, Sanders, & Hancock, 2013); physical presence (e.g. Kidd & Breazeal, 2004); human likeness (e.g. Bartneck, Bleeker, Bun, Fens, & Riet, 2010; L. U. Ellis et al., 2005; Hinds, Roberts, & Jones, 2004; Kiesler & Goetz, 2002; R. H. Kim, Moon, Choi, & Kwak, 2014; Kwak, 2014; von der Pütten & Kramer, 2012); animacy (e.g. Bartneck, Kanda, Mubin, & Al Mahmud, 2009); behavioral complexity (e.g. Rau, Li, & Liu, 2013; Scholl & Tremoulet, 2000; Vouloutsi, Grechuta, Lallée, & Verschure, 2014).

7 Selected examples: Intelligence attribution (e.g. H. M. Gray, Gray, & Wegner, 2007; Kiesler & Goetz, 2002; Sung, Guo, Grinter, & Christensen, 2007); mind perception/mind attribution (e.g. Epley, Waytz, & Cacioppo, 2007; H. M. Gray et al., 2007; Kamide,

Studies often try to explore the effect of different variables on the "success" of human-robot interaction by measuring participants' "attitude towards robots", "willingness to use robots", or "acceptance of robots".[8] In "an attempt to design and control not only robotics systems but also the entire process of human-robot interaction, users' performance included" (Zawieska, 2015, p. 3) these insights are meant to help with the improvement of future human-robot-interaction. Not surprisingly for such a complex issue, and considering the jumble of different variables, few widely accepted theories or models have emerged so far. One exception is Nicholas Epley and colleagues' (2007) Three-Factor Theory of Anthropomorphism, which strives to integrate the various perspectives investigated in previous studies, and suggests three "psychological determinants" of anthropomorphism: the accessibility and applicability of anthropocentric knowledge, the motivation to explain and understand the behavior of other agents, and the desire for social contact and affiliation. While Epley and colleagues explicitly named robotics as an area of application, and the model is frequently referenced in the HRI literature, they are not HRI researchers, but cognitive scientists. The success of their model in the HRI community demonstrates the close connection of HRI and the cognitive sciences in this particular context (which Section 2.3 will explore in depth).

Two basic assumptions underlie many of the HRI studies trying to find predictors for anthropomorphic attributions to robots: Firstly, the assumption that it is possible to "switch on" anthropomorphic attributions with the right kind of robot design or robot behavior. Secondly, the assumption that anthropomorphic attributions to robots are desirable and advantageous for human-robot-interaction.

Eyssel, & Arai, 2013); perceived social presence or sociability (e.g. Choi, Kim, & Kwak, 2014; Kiesler & Goetz, 2002; R. H. Kim et al., 2014; Schermerhorn et al., 2008); embarrassment (e.g. Bartneck et al., 2010; Choi et al., 2014); empathy (e.g. Darling, Nandy, & Breazeal, 2015; Riek, Rabinowitch, Chakrabarti, & Robinson, 2009a, 2009b; A. M. Rosenthal-von der Pütten, Krämer, Hoffmann, Sobieraj, & Eimler, 2013); hesitation to switch off/harm/"kill" robot (e.g. Bartneck, van der Hoek, Mubin, & Al Mahmud, 2007; Riek et al., 2009b, 2009a; Darling, 2012; A. M. Rosenthal-von der Pütten et al., 2013; Darling et al., 2015;).

8 Selected examples: Attitude towards/willingness to use/acceptance of robots (e.g. De Graaf & Ben Allouch, 2013; Kwak, Kim, & Choi, 2014; T. Nomura et al., 2008; T. Nomura, Kanda, Suzuki, Yamada, & Kato, 2009; Stafford, MacDonald, Jayawardena, Wegner, & Broadbent, 2014).

The first assumption is based on the notion that certain characteristics of robot technology – such as embodiment, mobility, autonomous behavior, or humanoid design – trigger a human perception system highly primed to recognize animacy (cf. Section 2.3). Commonly, "the robot is an inanimate object" is understood to be the default interpretation or null hypothesis. HRI studies then try to trigger anthropomorphic attributions in a controlled way, by manipulating the design or behavior of the robot. The idea behind this approach is that the existence or magnitude of certain features is able to push the robot over a "social threshold", giving it a "social presence" (Damiano & Dumouchel, 2018; cf. Levillain & Zibetti, 2017).

Few studies explicitly look at the opposite, at attributions of inanimacy. Presumably, it being considered the null hypothesis, this attributive perspective is not viewed as an interesting phenomenon per se. An exception to this is a cluster of research working with the concept of "dehumanization". Here, the idea is that "looking at a process of depriving objectified humans of characteristics regarded as crucial in order to be perceived and treated as a human" would contribute to "identify[ing] the key characteristics for robots to affect their anthropomorphism" (Złotowski et al., 2018, pp. 1 & 2; Waytz, Epley, & Cacioppo, 2010; Morera et al., 2018; cf. Haslam, 2006).

There are also some field studies finding anecdotal evidence of what we will also observe in the context of this book: robots being simultaneously enacted as an agent and as a thing, as both animate and inanimate. A study exploring spontaneous interactions with a social robot in a classroom setting found that for their study participants "seemingly contradictory features – a thing and a living creature – unproblematically coexist[ed]", "the robot present[ing] its multiple facets so that each theme c[ould] resurface at any particular moment" (Alač, 2016, pp. 12 & 15). A short field study with a hospital delivery robot found that the hospital staff perceived the robot as both a machine and a colleague, both "perspectives mutually coexist[ing], even for the same person" (Ljungblad et al., 2012, p. 9).

Studies like these, which consider attributions of both animacy and inanimacy, are rare, however. This might in part be due to the second assumption underlying many HRI studies: That of anthropomorphic attributions being beneficial for smooth human-robot interaction. They are thought to "facilitate … human-machine interaction, … increase people's willingness to care about the well-being of robots" (Złotowski et al., 2015, p. 351), and even to "facilitate … the introduction of robots in the society at large" (Ferrari, 2015, p. 17). This idea can also be encountered outside of a purely academic context.

For example, in the marketing campaign for the personal service robot Jibo, roboticist Cynthia Breazeal argued that "it is really important for technology to be humanized" (cited in Markoff, 2014). This assumption inspires HRI research efforts with an openly communicated agenda: In HRI, anthropomorphism is mainly studied in order to use the insights for the improvement of future robots' interaction capabilities and "usefulness ... by creating social bonds that increase a sense of social connection" (Epley et al., 2007, p. 897). In this context, knowledge about anthropomorphism is now "highly valued by many roboticists and computer scientists" for its potential to be used as a means to control user reactions to robots (Vidal, 2007, p. 3). Research efforts exploring the processes behind anthropomorphism are therefore frequently fueled by the inherent goal of building socially interactive robots:

> "While anthropomorphism is clearly a very complex notion, it intuitively provides us with very powerful physical and social features that will no doubt be implemented to a greater extent in social robotics research in the near future." (Duffy, 2002, p. 5)

In some parts of the robotics community, building a "perfect" human-like robot is considered the ultimate goal, or even "holy grail" (e.g. Duffy, 2006, p. 33): "It seems a truth universally acknowledged that a roboticist with a good research lab must want to create a humanoid!"[9] (Keay, 2011, p. 66). This goal is also fueled by the tempting engineering challenge it poses. As one of the roboticists interviewed for this book (cf. Chapter 3) explained:

> "[A humanoid robot] is not only the most extreme form [of robot]; it is the most difficult form of autonomous systems you can work on. ... It's the interactions, the possibilities of interaction ... they basically explode." (R2.3-00:07:22-8)[10]

The challenge is not only to make a robot look like a human but, even more, to make it behave like a human. While it is relatively easy to put a realistic looking "skin" on a robot "skeleton", making a robot autonomously move and speak in a completely natural-appearing way is still an unsolved problem.

---

9 A reference to the famous first sentence of Jane Austen's "Pride and Prejudice" (1813): "It is a truth universally acknowledged, that a single man in possession of a good fortune must be in want of a wife."

10 The numbers after this quote refer to the position in the audio transcript of the interview.

Robots with relatively good interaction abilities are usually remote controlled or follow a predetermined script, rather than acting autonomously – such as Hanson Robotics' Sophia[11], or Hiroshi Ishiguro's various Geminoids[12]. However, a realistic humanoid design is not a prerequisite for humans to experience a strong social connection and to attribute emotions, desires, and even personality traits to robot technology (cf. Chapter 1).[13]

The strongly innovation-driven goal of many HRI projects – to find out which robot characteristics have to meet which human characteristics in order to attain the best possible "interaction experience" – faces criticism from both within and outside the HRI community. Not only is there a range of methodological problems, such as the questionable operationalization of complex concepts like "human-likeness". There is also no clear evidence that human likeness actually has a positive influence on human-robot interaction.

HRI researchers face a methodological challenge: On the one hand, studies exploring human-robot interaction are supposed to use "realistic" scenarios in order to make the results generalizable and maybe even usable for marketable applications. On the other hand, variables like robot and environment features or user reactions need to be measurable and comparable. The result is often a methodological compromise, with research being conducted in laboratory environments with simulated "real life" scenarios, and metrics constructed around what is doable within the constrictions of the institutional conditions and available resources (Meister, 2014, p. 120). Often, this results in the use of "i-method"-approaches[14], as well as a naïve and uncritical use of what is understood to be "social science" methods by untrained engineers (Irfan et al., 2018). This leads to a lack of common metrics, methods, and generalizability – making the findings of most HRI studies neither comparable to each other, nor generalizable to a real life environment or a wider population (Steinfeld et al., 2006; Dautenhahn, 2007; Bethel & Murphy, 2010; Bischof, 2015).

These operationalization issues are also present in research on anthropomorphism and related phenomena in the context of robotics. For example, there is no consensus on what "human-like" robot design or robot behavior

11 https://www.hansonrobotics.com/sophia (accessed 2019-11-19).

12 https://eng.irl.sys.es.osaka-u.ac.jp/robot (accessed 2019-11-19).

13 E.g. J. Carpenter, 2016; Julia Fink, 2014; Forlizzi & DiSalvo, 2006; Kolb, 2012; Levillain & Zibetti, 2017; Sandry, 2015b; Sung et al., 2007; Yang et al., 2015.

14 Implying a designer's reliance on personal experience, attempting to take on a layperson's perspective (cf. Oudshoorn, Rommes, & Stienstra, 2004).

means, or how to "measure" anthropomorphism and users' attributions of human-likeness to robots – making study results difficult or even impossible to compare. Those difficulties are observable in arbitrary categorizations of robot designs in many studies trying to contrast "humanoid" against "non-humanoid" robots, with every study drawing the border between the categories somewhere else. There have been proposals for universal categorizations, such as Brian Duffy's (2003b) "Anthropomorphism Design Space for Robot Heads", but most HRI studies use "homemade" categories. Sometimes these border on the absurd, like when a robot vacuum cleaner with googly eye stickers is categorized as "human oriented"[15] (Kwak, 2014), or when an oven with arms is supposed to be "anthropomorphic" (Osawa, Mukai, & Imai, 2007). For one end of the design spectrum, there is at least a term most agree on – "humanoid". For the other end, a plethora of terms is in use, including "non-humanoid", "mechanistic", "mechanoid", "mechanical", "appearance-constrained", "single purpose", "functional", and "with few anthropomorphic features".

There is also no generally accepted measure for the "strength" of users' anthropomorphic attributions to robots. Most HRI studies do not operationalize this at all, but instead directly investigate the influence of different "human-likeness" levels on users' attributions of "mind", "sociability", or "intimacy", or emotional and behavioral reactions like empathy (e.g. Carpenter 2013; Garreau 2007; Garber 2013; Riek et al. 2009), embarrassment (e.g. Choi et al. 2014; Bartneck 2010), or decision making (e.g. Bartneck et al. 2007; Chandler & Schwarz 2010).

The overall methodological disunity is a topic of discussion within the HRI community. There are efforts for finding some consensus and comparability, for example by trying to make anthropomorphism "measurable" on a one- or two-dimensional scale (Bethel & Murphy, 2010; Ruijten et al., 2019), or by developing standardized tests for anthropomorphic attributions to robots (e.g. Bartneck et al., 2009; cf. Murphy & Schreckenghost, 2013). However, no generally accepted approach has been agreed on yet.

HRI researchers not only disagree on how to measure a robot's human-likeness and users' reactions to it. There is also no consensus on whether making a robot human-like is actually desirable. While there has been a steady stream of social robotics research based on the assumption that giving a robot the "right set" of lifelike features will somehow make users able and willing to

15 The vacuum cleaner without googly eyes, meanwhile, was categorized as "product oriented" (Kwak, 2014).

interact with the robot, and the construction of a "realistic" humanoid robot is considered by some as the "holy grail" of robotics, there is actually no consent within the robotics and HRI community on whether making a robot as humanlike as possible is worthwhile. In the context of this discussion, the so-called "Uncanny Valley" effect is referenced frequently. First proposed by Japanese roboticist Masahiro Mori (1970), the concept hypothesizes that the relation of a robot's human-likeness and observers' emotional responses is not linear. The underlying idea is that the more a robot is designed to look and behave like a human, the more positive observers react to it. With one crucial exception: If a certain level of human-likeness is reached – the robot resembling a real human very closely, but falling short of being a perfect representation – observers' reactions are adverse, even disgusted. The sharp dip in the graph representing the relationship of human-likeness and observer reactions is referred to as the Uncanny Valley. Although there is no clear empirical support for the hypothesis, even after decades of research, it is referenced frequently in the HRI literature (Brenton et al., 2005; Bartneck et al., 2009; Damiano & Dumouchel, 2018; MacDorman, 2019). However, the discussion surrounding the validity of the Uncanny Valley concept does direct attention to one important issue, namely that of the expectations a human-like robot raises with users. One of several explanations put forward for the (presumed) Uncanny Valley effect is that a very human-like robot design causes human users to have certain expectations about the robot's behavior, such as realistic movements or a smooth natural language interaction. At the current state of technology however, no humanoid robot is able to fulfill these expectations to a satisfactory level and – so the idea – the ensuing disappointment, irritation, or even disgust experienced by the user causes the Uncanny Valley effect (Ferrari, 2015; Złotowski et al., 2015).

Belief vs. Make-Believe

One profound issue is often overlooked in the discussion of users' expectations of human-like robots and the operationalization of their attributions of animacy to robots: That of whether users' behavioral and linguistic expressions of animacy attributions are founded in an actual belief that the robot in question is animate, maybe even driven by human-like intentions, or whether these expressions are merely metaphorical ascriptions, a performance of "make-believe", of "as-if the robot were alive".

We can find this distinction in several theoretical approaches to the attribution of animacy, agency, and intentionality to technological artifacts (Section 2.2 will discuss these terms in depth). John Searle (1983), for example, distinguished "intrinsic intentionality", which is based on existing mental states of a conscious living being, from ascribed "as-if intentionality", which is used in a metaphorical way to explain the actions of inanimate objects. Similarly, Epley and colleagues (2007) distinguished between "strong" and "weak" anthropomorphism. "Strong anthropomorphism" would entail the explicit belief that a nonhuman entity has humanlike characteristics, for example in the context of religious belief. In contrast, the metaphorical ascription of human likeness to artifacts known to be inanimate would be a form of "weak anthropomorphism". Eleanor Sandry (2015a, p. 11) used the term "tempered anthropomorphism" in a similar vein, meaning the "human understanding ... of the robot as somewhat humanlike or animal-like, but ... continually tempered by also perceiving the robot as a machine". Other authors propose that anthropomorphism can be understood as a spectrum with different shades or levels (e.g. Persson, Laaksolahti, & Lönnqvist, 2000).

Empirical studies in HRI, HCI (human-computer interaction) and HMI (human-machine interaction) research sometimes make distinctions like these. For example, the widely cited Media Equation study observed that users "mindlessly" attributed social attributes to computers – but also explicitly noted that none of the participants actually said that a computer should be understood in human terms or treated as a person (Reeves & Nass, 1996; also see Nass & Moon, 2000). The authors thus carefully ruled out anthropomorphism as a term to be applied to their observations. In the context of human-robot interaction, Leila Takayama (2012) observed different "levels" of anthropomorphic attributions being applied to the same nonhuman artifact and thus proposed to distinguish observers' "in-the-moment" perspective on robots from a "reflective" perspective. In the actual moment of interaction, a user might be quick to perceive a robot's behavior as agentic or even animate – a "visceral" interpretation, which can differ substantially from a more reflective perspective that would explain the robots behavior with the robot's programming.

Most studies do in fact refrain from operationalizing, or even just addressing, the complex, multifaceted nature of anthropomorphic attributions. This draws criticism from within the HRI community:

> "In the large body of experimental work on human reactions to anthropomorphic robots, responses on standard questionnaires are commonly taken to demonstrate that subjects identify a robot's displays or movements as ... expressions of the fundamental human emotions. ... Taking these responses ... at face value ignores the possibility that they are elliptical for the subjects' actual views. ... Saying that the robot has a 'happy' expression might be shorthand for the claim (for example) that if the robot were a human, it would have a happy expression." (Złotowski et al., 2015, p. 348)

The research discussed above shows that "metaphors that might represent a very weak form of anthropomorphism can still have a powerful impact on behavior" (Epley et al., 2007, p. 867), and that the power of "weak" anthropomorphism, of the "merely" linguistic and metaphorical attributions of animacy to technical artifacts, should not be underestimated. In scenarios of human-robot interaction, humans' ability to temporarily suspend their disbelief (Duffy & Zawieska, 2012) or even simply to "perform" the belief of a robot's animacy (McGonigal, 2003; cf. Jacobsson, 2009) can serve as a crucial facilitator for a smooth interaction. Anthropomorphic metaphors can serve as linguistic devices allowing efficient communication about technological artifacts – a "convenient fiction ... that permit[s] 'business as usual'" (Caporael, 1986, p. 218). This is especially relevant for complex and difficult to grasp technologies:

> "To confront the relatively unknown in an infinitely complex reality, we must rely upon our understanding of the relatively familiar. The resulting metaphorical concepts help organize inquiry and interpretation – they are necessary [and] fruitful." (Krementsov & Todes, 1991, p. 68)

We are very well able to understand metaphors as what Paul Ricoeur (1978, 2003) called "split reference", interpreting them simultaneously in a literal way, and as an imaginative concept. Nonetheless, it remains difficult to distinguish clearly between the playful, even useful, use of metaphors, the suspension of disbelief as an enabler for smooth human-robot interaction, and potentially harmful misunderstandings about the actual animacy of a technological artifact. After all, "every metaphor is the tip of a submerged model" (Black, 1979; cited in Watt, 1997, p. 60), and talking about a robot as if it were alive might correspond to having a mental model of a robot being a living being.

Are robot designers therefore guilty of deceiving users when they give robots human-like characteristics, when "robots are designed in such a way that they trigger us to 'fool ourselves'" (Turkle, 2011a, p. 20)? This question has been raised by several actors in the HMI and HRI community (e.g. Borenstein & Arkin, 2019; Coeckelbergh, 2018; De Graaf, 2016; Scheutz, 2012; Sparrow, 2002; Sparrow & Sparrow, 2006). Karolina Zawieska (2015) argues that the core of anthropomorphism is illusion and the topic therefore intrinsically tied to ethical concerns:

> "The main ethical issue lies not in deception itself but rather in a particular view of man where human beings are seen as creatures whose anthropomorphic projections can be evoked 'automatically' and their interaction with robots fully managed and controlled." (Zawieska, 2015, p. 1)

We will also encounter this discussion of deception, in varying forms, at the stops of our empirical tour along the life cycle of robots in the following chapters, and will revisit it once again the final discussion in Chapter 6.

In conclusion, animacy attributions – for example in the form of "anthropomorphic projections" – are a complex and controversially discussed issue in the HRI community. In the context of HRI studies, the focus of academic interest is almost exclusively on the actual or potential interaction between a robot and a human user. However, this moment of interaction is only one very narrow "slice" of the whole life cycle of robots. In the following chapters, we will see that animacy attribution is also an influential phenomenon in all other stages of the cycle. In three exemplary explorations – of robotics engineering practice, of demonstrations, science communication and marketing, and of media discourse on robotics – we will encounter different forms of animacy attribution, and explore its context-specific constructive role.

2.2. Terminology: Anthropomorphism, Agency, Animacy, and More

Before we continue our tour along the life cycle of robots, we first must clarify some of the terminology used in this book. This section will tease apart several overlapping concepts – such as animacy, agency, and intentionality – and it will establish "attribution of animacy" as a central term for this book.

In the vast body of scientific literature on human-robot interaction (cf. Section 2.1) the term used most often for the phenomenon of humans ascribing lifelike qualities to robot technology is "anthropomorphism" – meaning "the attribution of human traits, emotions, or intentions to non-human entities" (OED, n.d.-d). Its derivation from the Greek "ánthrōpos" ("human") and "morphē" ("form") points to a crucial limitation of the term. By definition, it refers to the attribution of human characteristics to something. In the context of robotics and HRI, however, anthropomorphism is often used to mean something else. Firstly, the term is often (mis)used to describe the human-like design or behavior of a robot, instead of the phenomenon of attribution (Julia Fink, 2014, p. 63; cf. Bartneck & Forlizzi, 2004). This disregards that a robot "is not anthropomorphic per se, but only in so far as it gives rise to anthropomorphic processes in a given user and situation" (Persson et al., 2000, p. 1). Secondly, the term anthropomorphism is frequently used to describe a much wider phenomenon: the attribution of characteristics of living beings in general to robots. Characteristics such as aliveness, emotionality, personality, and sociality are not unique to humans, but apply to a much wider group of living entities. Rarely, the term "zoomorphism" is used for the attribution of characteristics of nonhuman animals to robots. A "zoomorphic robot" is usually understood as a robot with an animal-like design.

In the following chapters, we will encounter several instances of features being attributed to robots that sometimes are characteristic to living beings in general (such as sensory experiences, intentions, or emotions) and sometimes are more specific to human beings (such as long-term life goals). In some existing concepts, this phenomenon is understood to be one level of anthropomorphism (e.g. Persson et al., 2000). But for the purpose of this book and the phenomena it describes the wider term "attribution of animacy" is more adequate.

"Animacy" is a grammatical and semantic feature meaning the "the quality or condition of being alive or animate" (OED, n.d.-a), the adjective "animate" meaning "endowed with life, living, alive" (OED, n.d.-b). Their antonyms "inanimacy" and "inanimate" will also play a role in this book. Animacy also happens to be used in the cognitive sciences and developmental psychology in the context of research exploring, for example, the perceptual and attentional processes involved the identification of living entities in our visual environment (cf. Section 2.3). With research in HRI drawing heavily on the cognitive sciences (cf. Section 2.1) the term animacy also made its way into the robotics literature. However, animacy, with its connection to animism, comes with a difficult colonialist connotation, which is rarely discussed reflexively, or even acknowledged, in the HRI and cognitive science literature (cf. Section 2.3).

Despite this connotation, this book will use the term "attribution of animacy" for the phenomenon in the focus of interest – for two reasons: Firstly, "animacy" is used in the majority of the relevant HRI literature. Another possibly adequate term – "aliveness" – has only been used by a handful of authors (e.g. Turkle, 2010; Sandry, 2018). Secondly, "attribution of animacy" can be understood as something like the lowest common denominator of the different variations of the phenomenon this book explores. A more confined term like "anthropomorphism" would not adequately reflect the observation that people also ascribe physiological processes – which are not unique to humans – to robots.

The term "attribution of agency", too, would be too restrictive for the context of this book. It is, however, important to acknowledge the importance and relevance of the concept of agency. Depending on the disciplinary context (sociology, philosophy, cognitive sciences ...), definitions of agency focus on slightly different aspects. At the most basic level it is "the at least partially independent capacity to engage in goal-directed action" (H. M. Gray et al., 2007).

At this point, it is important to note that

> "the concepts of 'animacy' and 'agency' ... are not coextensive. Animate entities are living things that can act as agents .... Living things that are not sentient and do not act as agents, such as trees and mushrooms, are not animate. The domain of agents, however, can include inanimate automatons, such as robots, that generate their movements and actions to achieve goals." (Gobbini et al., 2011, p. 1911)

Science and technology scholars have long been discussing whether non-biological entities can possess agency (also see Section 2.3). Werner Rammert (2008) proposed a multi-level model of agency. On the model's lowest level (causality), agency means simply "behavior that exerts influence or has effects". This level, on which "it doesn't make any difference whether humans, machines or programs execute the action" (Rammert, 2008, p. 11), has obvious parallels to the concept of generalized symmetry within Actor Network Theory: "Objects too have agency" (Latour, 2005, p. 63). The next higher level (contingency) requires the capacity to act differently, to choose among several behavioral options. Only the third level uses the term intentionality, referring to reflexive and intentional actions. Rammert (2008, p. 12) argued that technical artifacts, while not able to have "literal" intentionality, "can be constructed as if they had an intentional structure". Chapter 4 (Section 4.7) will explore in more depth the issue of technical artifacts acting "as if" they were animate.

In the cognitive sciences and HMI literature one can find many more proposals for the conceptual relations of the terms discussed in this section. For example, Heather Gray and colleagues (2007) understand agency – defined here as the capability to act and intend – as one dimension of "mind" that can be attributed to an agent or entity (next to the dimension of experience, i.e. the capability for feelings and sensations). Elsewhere, Florent Levillain and Elisabetta Zibetti (2017, p. 13) propose that the behavioral cues of autonomously acting technological artifacts are interpreted by an observer on three levels: the Animacy Level (Does the object look alive?), the Agency Level (Does the object appear to act intentionally?), and the Mental Agency Level (Does the object appear to take into account others' goals?).

The terms discussed above (anthropomorphism, animacy, animism, agency, intentionality...) are those one encounters most frequently in the current academic literature. There are also less frequently used concepts, such as "Universal Projection" – used by Thomas Luckmann (1983) to describe humans' capacity to project their own living body onto everything they encounter in the world, which is sometimes referenced in the context of HRI (e.g. Nørskov, 2017, p. 11). As is "Mythopoeic Thought", a proposed ancient form of human thought, in which each observed event is attributed to the will of a personal being (Frankfort et al., 1946; referenced e.g. by J. Carpenter, 2016, p. 20).

2.3. Disciplinary Perspectives: Animacy Attribution as an Object of Research vs. Methodological Malpractice

While HRI is the academic field where most research on animacy attributions takes place at the moment, the issue is also of interest for many other disciplines. An exploration of the publications of different academic fields reveals two overarching perspectives: Firstly, animacy attributions as a methodological malpractice, and secondly, as an object of research in itself.

For centuries, scientists freely compared natural phenomena to processes of the human body and mind. Medieval scholars attributed chastity to camels and self-sacrifice to storks, renaissance scholars referred to nature as a benevolent servant or artist (Daston, 2000, p. 29). By the seventeenth century, however, natural philosophers started to abandon these comparisons. The explaining of natural processes with human-like beliefs and desires became to be considered scientific misconduct: "Nature had become irretrievably 'the other'" (Daston, 1995, p. 38, cf. 2000). It took a while for this perspective to reach the non-academic community. Up until the nineteenth century, fed by the reports of travelers, naturalists, and amateur scientists, zoopsychological publications describing animal behavior with "human" terms stayed wildly popular, "replete with descriptions of 'states' and 'factories', 'art' and 'crafts', ... 'friendship', 'wars' ... among animals" (Krementsov & Todes, 1991, p. 76). By the end of the nineteenth century, criticism of the zoopsychological perspective on animals reemerged and most scholars agreed that "in no case may we interpret an action as the exercise of a higher psychical faculty, if it can be interpreted as the outcome of the exercise of one which stands lower in the psychological scale" (Morgan, 1894, p. 53). In the 1960s, researchers studying great apes – among them Jane Goodall – were strongly criticized for attributing presumably human characteristics, such as emotions, to animals – the "worst of ethological sins" (Goodall, 1993, p. 15; cf. Rees, 2001).

Animism – the attribution of life or spirit to nonliving entities (OED, n.d.-c) – was traditionally viewed as an immature disposition. As such, it stayed of interest mainly in two academic contexts: On the one hand, research in developmental psychology on certain phases of infants' cognitive development (e.g. Piaget, 1929). On the other hand, early anthropological research on "primitive" religions ascribing a distinct spiritual essence to objects, places, and creatures (cf. Franke, 2010). This "old animism" perspective of anthropology, viewing "primitive" animist cultures as being unable to differentiate between persons and things, was held, for example, by nineteenth-century anthropologist Edward Burnett Tylor (e.g. 1871). Today it is criticized for its colonialist and dualist worldviews and rhetoric (Harvey, 2006, p. xii). Nonetheless: "images of fetishes, totems, ... tribal art, pre-modern rituals, and savagery ... have forever left their imprint on the term [animism]" (Franke, 2010, p. 11). Up until recently, practices of attribution of animacy to non-living entities were regarded in most scientific disciplines as both an archaic or infantile reflex and as a methodological mistake (Vidal, 2007, p. 3). As something that scientists knew having to avoid at all costs for its "violat[ion] of the ideal of the objectivity of perspective"[16] (Daston, 2000, p. 28), little scientific attention was directed to the nature and consequences of animacy attributions for a long time:

16 Translated from German by the author.

> "The debate about the nature and implications of anthropomorphism has rarely been neutral or scientifically objective but has focused mainly on its fallacious essence ... which has diverted attention away from the goal of understanding the nature of the phenomenon." (Urquiza-Haas & Kotrschal, 2015, p. 168)

Only in the last few decades, fueled by observations of human interactions with increasingly complex and autonomous technologies, scientific interest reemerged across academic disciplines (cf. Vidal, 2007). For example, communication scientist Sherry Turkle, based on her ethnographic research on computer users, proposed that computers were more than "just a tool" and explored how we interact socially with them (Turkle, 2005, p. 3). She would later coin the term "evocative objects", describing how certain machines "can act as a projection of part of the self, a mirror of the mind" (ibid., p. 20) and can even become emotional and intellectual "life companions" (Turkle, 2011b, p. 9). Similarly, communication scientists Byron Reeves and Clifford Nass showed with their Computers as Social Actors paradigm that even minimal social cues from a technical artifact can cause humans to mindlessly treat it like a living interaction partner (e.g. Nass et al., 1993). Their observations are often referred to as the Media Equation, after the title of their widely cited book (Reeves & Nass, 1996). In a series of HCI studies they showed that human "individuals are responding mindlessly to computers to the extent that they apply social scripts – scripts for human-human interaction – that are inappropriate for human computer interaction" (Nass & Moon, 2000, p. 83). Communication scientist Don Ihde (1990, p. 97 ff.) explored the interaction of humans and machines as a quasi-other, proposing "alterity relations" as a term for relations with technology (cf. Sandry, 2018).

In the cognition science community, a widely cited study from 1944 by Fritz Heider and Marianne Simmel showed that human subjects interpreted movements of abstract shapes in an animation film as social interactions between animate entities. For decades, this study was mainly perceived as an interesting anecdote (Aarts, Dijksterhuis, & Dik, 2013). In recent years, however, it has been replicated several times and is now regarded as seminal for the research of social perception and causal attribution (cf. Lück, 2006). Today, there is a lively research community interested in the cognitive processes and neural structures involved in the perception of animacy and action – both in the developing and adult brain (e.g. Gobbini et al., 2011; Marsh et al., 2010). The ability to identify animate entities is already present in infants

(Kuhlmeier, Wynn, & Bloom, 2003; Woodward, 1999). It is understood to be the foundation for the later development of a Theory of Mind – the ability to attribute internal mental states to others (Premack & Woodruff, 1978). Research in the cognitive sciences found that agentic entities in our visual field are prioritized via attentional selection, compared to inanimate objects (e.g. New, Cosmides, & Tooby, 2007; Scholl & Gao, 2013). Which entity is categorized as animate depends not only on the visual appearance (e.g. the presence of eyes), but also on its behavior. For example, the perception of movement being goal-directed and self-propelled strongly contributes to an entity being categorized as behaving intentionally and having human-like mental states, and to the observer behaving towards the entity as if it was alive (see e.g. Epley & Waytz, 2010, for an overview). John Harris and Ehud Sharlin (2011) explored human reactions to abstract motion with the help of an extremely minimalistic robot consisting of nothing but a stick, which was remote-controlled to perform different movements. Observers not only consistently rated certain movements as emotional expressions (e.g. speed – excitement, approach – aggression), but also spontaneously tried to find meaning in the movements and attributed mental processes to the robot.

A proposed explanation for these reactions is that the cognitive-perceptual subsystem responsible for the identification of agentic entities in our environment is so sensitive that it is prone to over-interpret even minimal perceptual cues. The evolutionary reasoning is that erring in favor of interpreting an object in our environment as animate increases the probability for survival. From an evolutionary perspective, being able to detect other intentionally acting agents in our vicinity is a crucial fitness advantage. Being able to quickly identify a predator can mean the difference between life and death (e.g. B. J. Ellis & Bjorklund, 2005). Also beyond the immediate threat of being killed by a wild animal, humans, as highly social animals, have been profiting from this ability in the context of their complex social lives – for example when establishing alliances with other human tribes. This idea was conceptualized in the so-called Social Intelligence Hypothesis (Kummer et al., 1997). The idea of a "Hyperactive Agency Detection Device" is even proposed as an explanation for religious beliefs in a higher power (Barrett & Lanman, 2008):

> "Based on stimuli in the moment, we ascribe the highest level of sophistication possible to the object at hand. … The smallest evidence of live or intentional action encourages perceptional shift, allowing us to ascribe live and intentional statuses to objects more readily." (Owens, 2007, p. 573)

Animacy as a default interpretation of ambiguous stimuli has been proposed by several researchers. For example, Daniel Dennet (1998) postulated that the "Intentional Stance" is the most abstract of three possible levels of abstraction[17] when considering the mental state of an entity. When taking the Intentional Stance, predictions made for the behavior of an entity are based on its assumed beliefs and desires – compared to, for example its physical properties, respectively its design purpose, on the two less abstract levels. Similarly, Stewart Guthrie (1997) proposed an "involuntary perceptual strategy", and Linnda Caporael and Cecilia Heyes (1997) a "cognitive default", in that "we will default to human characteristics whenever going gets rough" (ibid., p. 64). Within the cognitive sciences, the phenomenon of animacy attribution is considered "endemic" (Watt, 1997, p. 125), "almost irresistible" (Eddy, Gallup, & Povinelli, 1993, p. 88), and "inevitable" (Krementsov & Todes, 1991, p. 80), and researchers are trying to "set traps" (Caporael, 1986, p. 217) for it, in order to "tame" it for research and application development – as discussed for the context of HRI studies in Section 2.1 of the present chapter.

In the 1980s and 1990s, the field of science and technology studies (STS) began to explore the agentic and interactive role of technological artifacts. It was the context of scientific practice where STS researchers first began to explore the crucial impact of non-human artifacts – such as microbial samples or scientific instruments – on practices of scientific knowledge production (e.g. Knorr-Cetina, 1981; Latour & Woolgar, 1986; Lynch, 1985). This research resulted in a re-conceptualization of the prevailing ontological separation of "the social" and "the technical" into a concept of human and technical agency existing in parallel (e.g. Bijker & Law, 1992; Latour, 2005; Law, 1991; MacKenzie & Wajcman, 1999; also see Krummheuer, 2015). In a "turn to technology", researchers began exploring the social shaping and construction of technology (Woolgar, 1991). Rather than looking at the impact of technology on society, this research was – and still is – interested in how societal context finds expression in technological developments, exploring ideas such as material agency (cf. Knappett & Malafouris, 2008) and artificial interaction (e.g. Braun-Thürmann, 2002). Lucy Suchman (1987, 2007) described new human-machine configurations and human interaction with intelligent machines in her "Plans and Situated Actions". Karin Knorr-Cetina (1997, 1998) proposed the concept of a "sociality with objects" after observing human-object relationships with a perceived mutuality and solidarity.

17 Physical Stance, Design Stance, Intentional Stance.

Crucially, while this research on “autonomous technology [took] place down on earth … it also influence[d] the higher spheres of philosophical debates about the ideas of agency and autonomy” (Rammert, 2011, p. 1). Actor Network Theory (ANT) proposed a radical symmetry of human and nonhuman actors (“actants”), meaning that both are fundamentally equal in their contribution to any effects they have on the environment (cf. Section 2.2). In coming together in heterogeneous networks, human and nonhuman actants are presumed to constitute sociotechnical ensembles, which, as a whole, serve as the location of any agency and create meaning in the world (e.g. Latour, 1987, 2005; Callon, 1986). In contrast to ANT’s approach, Werner Rammert and Ingo Schultz-Schäffer (cf. Section 2.2) suggested a distribution of agency between humans and technical artifacts, with the attribution of agency to human or nonhuman agents being constructed only by the observer (Rammert, 2002, 2008, 2011; Rammert & Schulz-Schaeffer, 2002a).

Already in the next chapter, we will encounter such a perceived distribution of agency – between roboticists and “their” robots. Also in the following chapters, while exploring a range of practical and discursive human-robot interaction, we will revisit and apply many of the conceptual approaches discussed above.

3. Making Robots: In/Animacy Attributions in Robotics Research and Development

3.1. Complex Epistemic Practices in Long-Term HRI

Contrary to what countless headlines in the news suggest, robots are not coming to us from the future or stepping out of science fiction movies (cf. Chapter 5). All real-life robots – from the huge industrial robot arms hidden in factories to the small vacuum cleaning robots in our households – were developed and constructed by human researchers, engineers, and workers.

Robotics research and development (R&D) takes place along a spectrum of private, academic, and commercial contexts: From small-scale development projects run by one person to huge robotics institutes with hundreds of employees; from tiny startups to big industrial players with dedicated robotics R&D departments; from hobby inventors and nonprofit organizations, academic research institutions and public-private partnerships, to purely profit-oriented businesses. In these environments, roboticists build robotic hard- and software from scratch or by combining existing components. They develop new robots, assess and improve features of existing robots, and deploy them in new application scenarios.

This chapter makes an empirical stop on our tour along the life cycle of robots at one specific section of this spectrum of robotics R&D: the academic context. Without presuming that observations made in this one particular context can be generalized to the whole spectrum, this chapter's observations will nonetheless be able to give crucial insight in the unique relationship that robotics professionals have with the robots they develop, build, and work with.

The special focus of our interest is, of course, on whether and how robots are perceived and represented as quasi-animate entities in this particular context. One could assume that, due to their expert insight, roboticists would be less susceptible than non-experts to attributing characteristics of living beings to a robot. After all, contrary to lay users of robot technology, roboticists usually have expert technical knowledge about a robot's hardware setup, and know exactly which control algorithms are responsible for the robot's performance. The present chapter will counter this assumption by drawing on observations made in interviews with roboticists working in university robotics laboratories. It will show that these roboticists do in fact routinely attribute animacy to their robots. Crucially, they do not do this in the form of an inflexible, one-sided attribution, but by constantly switching between attributive perspectives on the robot. We will see that this practice of representing the robot as both an inanimate object and an animate being is an integral and constructive aspect of roboticists' work.

As discussed in Chapter 1, our question is not whether humans (here: roboticists) are correct in attributing inanimacy or animacy to robots. Instead, we will explore how, and with which discursive and material consequences, these attributions are enacted (cf. Suchman, 2007, p. 2). In this, we follow the academic tradition of the science and technology studies (STS), as well as ethnographic and discourse analytic "laboratory studies". It is not a surprising insight per se that the professional environment of robotics research and development is a setting for complex epistemic practices. We know from existing research in STS that scientists' interactions with technical artifacts are a crucial and constructive aspect of the knowledge production process. As discussed in Chapter 2 (Sections 2.2 and 2.3), it was on the basis of their observations of professional practices in scientific laboratories that STS scholars first articulated the crucial impact of technological and other nonhuman artifacts on practices of scientific knowledge production (e.g. Knorr-Cetina, 1981; Latour & Woolgar, 1986; Lynch, 1985). One central observation of this research was that the process of knowledge production is not a smooth, controlled process. On the contrary, everyday academic work was shown to be a mess of trial-and-error and tinkering (Knorr, 1979). With this in mind, we will approach the context of robotics R&D practices as an opportunity to learn how roboticists' interactions with the robot technology they employ and construct not only shapes the technology itself, but also the discourse about the technology – both within and outside of the R&D context.

This chapter will show how a R&D process dominated by constant experimenting, tweaking, customizing, and being confronted with unplanned results, contributes to roboticists' attributions of animacy to the robots they work with; how the constantly changing demands and challenges of roboticists' professional lives make their ability to playfully balance and switch between seemingly contradictory perspectives on a robot crucial and constructive.

A connection of the discussion of technical agency (via laboratory studies) to the laboratories where robots are developed appears to be quite obvious. After all, robotics is a technology featured heavily on the stage of the technical agency discussion (cf. Chapter 2, Section 2.2). Nonetheless, the particular case of roboticists' research and development practice has received little academic attention so far. One of the few studies looking explicitly at roboticists' work from an STS perspective is Andreas Bischof's (2015, 2017a) work on epistemic practices in social robotics. Bischof analyzes the practices and strategies employed by social robotics researchers and engineers in solving the "wicked problem" (Rittel & Webber, 1973) of deploying socially interactive robot technology into everyday contexts. The specific question of whether roboticists attribute animacy to their robots is only touched indirectly, however, insofar as Bischof observes roboticists' staging practices in the context of demonstrations and science communication efforts (which we will explore in Chapter 4).

Most other research with an interest in animacy attributions to robots – predominantly in the field of human-robot interaction (HRI) studies – focuses mainly on the interaction of lay users with robot technology (cf. Chapter 2, Section 2.1). Moreover, the empirical studies in the context of these research efforts usually take place in rather artificial laboratory environments, instead of observing spontaneous human-robot interaction in the field, and only look at very short periods of interaction. These constraints are primarily owed to the limited spread of robot technology in everyday environments. This means that there are only few opportunities for field research on long-term human-robot interaction. Interactions with the consumer robot applications available on the mass market today (mainly cleaning robots) could provide the opportunity to study long-term interactions in customers' everyday environment. Also the everyday work of machine operators working with large industrial robots, for example in the manufacturing industry, could serve as fields of research. However, the robots employed in these specific contexts are usually not considered socially complex enough to be of interest to HRI researchers. The underlying assumption being that non-socially interactive and non-humanoid robots do not provide enough opportunity for research-worthy interactions.

The widespread approach of researching only short-term interactions with non-expert users in controlled laboratory settings disregards a whole group of users, who do in fact have long-term and hands-on experience with robots: Roboticists – the engineers and researchers who develop the very object of interest. After all, with complex and interactive robot technology not yet being freely and affordably available to the average consumer, robotics laboratories are one of the few contexts where intensive long term-interaction with robots can be observed today.

An additional reason for the lack of scholarly attention to the robotics R&D context might be that "roboticist" is a relatively new profession. Few researchers and engineers who identify as roboticists today have a degree in robotics, as degree programs offering dedicated robotics training are a relatively new development. Most who work in robotics R&D today have an electrical or mechanical engineering, or computer science background. The heterogeneity of the whole discipline is also reflected in the composition of R&D teams, each roboticist bringing with them the practices of their original field(s) of training. At the same time, not every researcher or engineer working on robots will call themselves a roboticist. Reasons for this are the lack of consensus of what a robot actually is (cf. Chapter 1, Section 1.5) and the fields' wide overlaps with other disciplines, such as artificial intelligence and automation engineering.

3.2. Approach

Cases and Method

The focus of this chapter will be on observations made during semi-structured narrative interviews with eight roboticists (referred to in the text as R1–R8). The interviewees were employed as doctoral and postdoctoral researchers at robotics laboratories at the computer science, electrical engineering, and computer engineering departments of a large technical university in Germany. Despite its small size, this sample of cases was able to provide a range of perspectives and valuable insights into roboticists' R&D work in an academic context. At the time of the interview, half of the interviewees worked with robots that had some vague humanoid features, such as arm-like manipulators or a "head" at the highest position of the robot body. Two worked with "mechanical" looking robots like drones or small industrial robot arms. However, all of the interviewed roboticists had experience with more than just the robot type they currently worked with. They had encountered a range of robots in their earlier work, at conferences and trade fairs, and in their private lives. Their experiences with robotics R&D were not only based on the work at their current workplace, but also on their experiences at the various international robotics research institutions they worked at during earlier stages of their career.

All participants signed a standard declaration of consent and received no compensation for their participation. The interview sessions entailed relatively focused questions about the roboticists' everyday work practices with robot technology in general and specific robotic artifacts, as well as the discursive practices within their teams. Furthermore, the interviewees were encouraged to narrate freely about the relationship with "their" robots and to reflect on which role robot technology played for their professional lives. Four of the interviews were conducted in German; quotes from these interviews have been translated into English by the author of this book. The other four interviews were conducted in English. However, as English was not the first language of these interviewees, it was, in some cases, necessary to slightly edit direct quotes for the sake of comprehensibility. All edits are indicated by square brackets.

All interview transcripts were analyzed following a qualitative content-analytic approach (Mayring, 2010). Analytical categories were developed inductively and iteratively from the material, the central criterion being instances of discursive and non-discursive animacy attribution to robots in the wider sense (including attributions of physiology, sensory experience, cognitive processes, intentionality, sociality, personality, emotion), as well as hints to practices of staging robot agency and animacy (e.g. in the form of a purposeful backgrounding of remote controlling of robot activity).

Chapter Structure

The present chapter is structured along its main conclusions. First, it will focus on the central, cohesive role of the physical robot platform for robotics R&D teams (Section 3.3). Second, the chapter will show how robots' crucial role as feedback-givers in the R&D process contributes to them being perceived as valuable team members (Section 3.4). Third, we will see how robot's often unpredictable behavior during experimental periods of the R&D process act as a perceptive trigger for attributions of animacy (Section 3.5). Finally, the chapter will show that a constant switching of perspectives and attributions of in/animacy to robots is an integral and constructive aspect of roboticists' work (Section 3.6).

3.3. The Robot Body in the Center of Attention

Robotics researchers and engineers are often depicted in the media next to a robotic platform, sometimes with their arm draped around it in a companionable manner. In their actual work life, many roboticists spend the majority of their time in front of a computer. Most roboticists interviewed for this book pointed out that working with and on a physical robot platform is only one aspect of their work. Usually, a lengthy period of working in a software environment precedes any practical work with robot hardware. Only after a control algorithm has been developed to a certain point can it be implemented and tested on a physical robot platform. However, even during the period spent mostly at a computer and not (yet) physically near the actual platform, the embodied robot is always in a roboticist's focus of attention.

One reason for this is the robot's unifying role within a roboticist team. The term "robot platform" reflects this role: The robot serves as a shared platform for the team members' different tasks. As robotics is not a homogenous discipline with standardized terminology and methodology, researchers and engineers working in robotics laboratories usually come from a range of different backgrounds. Most trained as electrical, mechanical, or computer engineers. Increasingly there are also roboticists with a background in more "exotic" disciplines, such as cognitive psychology or user interaction design. Frequently, researchers from different academic backgrounds and with slightly different research interests cooperate in a robotics project. Therefore, "roboticists teams can have as many scientific and non-scientific goals as a robotic system has relevant components" (Bischof, 2015, p. 38; cf. Meister, 2011, p. 109), which can make cooperation and communication among the different team members a challenge. In this context, the robot platform provides a shared focus of attention and action – taking the role of what STS calls a "boundary object" (Star & Griesemer, 1989; cf. Bischof, 2015, p. 38). For example, R4's research project had dedicated roboticists working on separate components of one robot, "because everyone has their sub-field for which they are the expert" (R4.3-00:03:24-4[1]). Some developed pressure sensors for its "skin" surface, some worked on its navigation system, some on the control of its manipulators ("arms"), and some on the user interface. The common goal of developing separate robot components or algorithms and integrating them in a complex robot platform brought all team members together.

This is a typical situation for academic robotics R&D. In contrast to commercial robotics R&D, the majority of robotics projects in academia do not produce a finished, marketable robot platform. Instead, the goal is usually to either incrementally improve certain technical features, or to apply existing technological solutions in novel ways. Either way, there is usually an expectation to present the results in a way that highlights the platform's applicability to a real life problem. As discussed in Chapter 1 (Section 1.2), there is a strong, mainly politically driven, expectation to deploy robot technology in as many contexts as possible. This is in part fueled by an international "arms race" for shares in the rapidly expanding robotic market (Bischof, 2017a, p. 138): "Each time a robot acquires a new capability, a search for applications that can take advantage of that new capability follows" (E. S. Kim et al., 2012, p. 3). One consequence of this race is that the enormous funding sums poured into robotics research on the national and international level are often contractually tied to the development of demonstrable, preferably even marketable, robotic applications. In practice, this challenge of finding an impressive application is another factor keeping the physical robot at the team's center of attention.

Theoretically, it would be possible to develop and test robot control algorithms with a simulation of a robot in a virtual environment. However, this is only possible up to a certain point of complexity. For most roboticists, the final objective is to apply their work in the real world, to modify the physical environment with their robot. If this is the goal, then working with a physical robot platform is basically inevitable. R3 explained: "I use [the robot platform] because I need [it] to test the algorithms. I can test with synthetic data, but it's not the same" (R3.3-00:07:07-1). Even when the expected end result is not a fully marketable, deployable robotic platform, an underlying expectation remains that roboticists prove the success of their work with a physical demonstrator. For many, especially high-level, robotics conferences and journals experimental proof is even a prerequisite for publication. Here, demonstrations – usually in the form of laboratory experiments documented on video – are

1 The numbers after this and the following quotes refer to the position in the audio transcript of the interview.

required to show that the theoretical work described in a publication actually works in the real world. R3 explained that their institute's director insisted that "if [he] do[es]n't see it working he do[es]n't believe it" (R3.5-00:08:09-9). In Chapter 4, we will dive deeper into the context of robotics demonstrations, and explore how they employ attributions of animacy not only to prove a robots' functionality, but also to illustrate its applicability to desired use scenarios. For now, our focus shall remain on the work that roboticists do long before a demonstration.

In spite of the general expectation to produce demonstrable results, not all roboticists work with a physical robot. Two of the interviewed roboticists reported that in their specific sub-field it was acceptable to publish results of just a simulation study. However, they themselves were critical of this practice. R3 pointed out that "testing with synthetic data [is] not the same. [It] seems kind of fake, because you can tweak the data" (R3.3-00:07:07-1). Taking the step from a simulation environment into the real world poses a completely new set of challenges. R7 reported that their research group's simulation-only approach and avoidance of real-world experimental testing resulted in their robots not being ready to use by the end of the project (R7.1-00:07:52-9). While in a simulation the environment is controllable, physical reality comes with all kinds of interferences which can disrupt a robot's performance. We will come back to this phenomenon in Section 3.5.

The robot's status as a boundary object for the team is further amplified by the amount of physical work going into it. Affordable standardized robotics hard- and software has only relatively recently become available. Consequently, most roboticists are required to customize existing hard- and software, or even to create it from scratch. As Andreas Bischof (2015, p. 62) points out, the required time- and labor-intensive tinkering work (cf. Knorr, 1979) sometimes makes the final result somewhat of a bricolage (cf. Lévi-Strauss, Weightman, & Weightman, 1966). Indeed, also the robots used and built by the roboticists interviewed for this book were a conglomerate of commercially available modules and homemade components. Almost all of the interviewed roboticists reported to be involved in this building or customization process. While none of their platforms were built completely from scratch, all were in some way custom built from off-the-shelf modular parts. This modular approach is a fairly typical practice. Building a robot from scratch would consume too much time and individual modules are readily available from various manufacturers. These materials are costly, however, so their use depends on how much funding is available to the individual project. In order to save money, nonfunctional or unused platforms are often dismantled and the parts reused for new projects. R3, for example, described how budget restrictions in a former project forced them to laboriously build their hardware from scratch, resulting in less time being available for the actual scientific work.

It is here, with the roboticists' tinkering practices, where we encounter the first obvious instances of animacy attributions in the present chapter. When speaking about the construction process – be it building a robot from scratch or combining existing modular parts – the interviewed roboticists frequently referred to their robot platforms with terms usually reserved for living bodies. Even when the robots in question did not have a humanoid design, their individual parts were discussed as if they were biological body parts. R4 was quite aware of this, describing it as "projecting the [robot's] form on a humanoid form" (R4.3-00:07:43-4). The practice is partly rooted in a need for easy communication. When referring to the topmost part of a robot, the term most easily understood by everyone is simply "the head" – even when the robot does not have an explicitly humanoid design.

However, the practice goes beyond just communicative ease. By choosing the placement of components on the robot, roboticist can influence its final shape – and its resemblance to an animate being. One of the interviewed roboticists explained with obvious joy how they and their colleagues decided to make their – quite "mechanical" looking robot – more animate-appearing: When they had to place three antennas on the robot, they decided to install two of them on the sides of its "head", and one on the very back of its chassis. This ended up giving the robot "ears" and a "tail".

Projections of human physiology on the robot were also reflected in roboticist's discursive practices. Not only body parts, also physiological processes were attributed to robot platforms. Especially situations in which a robot did not function as intended were described with terms of illness and injury. R4 vividly described how their robot had a "heat stroke" (R4.2-00:00:55-3) due to its cooling system failing. The robot also suffered a "fracture[d bone]" when one of its wheels broke, causing it to "need a doctor" (R4.3-00:05:22-5). As we will explore in depth in Section 3.5 of this chapter, situations like these, in which the robot behaved in an unexpected way, were among the strongest triggers for animacy attributions to robots in this particular context.

The practice of customizing robot hardware, as described above, is another typical context for the attribution of physiological processes. When R1 explained how they used components of an older platform to improve a newer one, they described it as "cannibalizing" the old robot. It had to "offer its health for the good of the other" robot (R1.3-00:04:15-9). In one of the robotics laboratories visited for this book, there even was a communal robot "cemetery". On a dedicated shelf labeled "Nao[2] Cemetery", nonfunctioning robots and robot parts waited to be "dismembered" and "revitalized" as spare parts for newer robots.

3.4. The Robot as Tool and Team Member

In contrast to the references to the robot as a biological body described in the previous section, when asked directly which role the robot played for their work most interviewees spontaneously referred to it as a technical object. Several called it a research tool, explaining that it was used "for developing ... ideas" (R3.8-00:00:06-0) or "to evaluate ... computer models" (R2.2-00:01:54-0). A robot used as a demonstrator was a tool "to show ... research" (R8.1-00:06:38-8). Some roboticists – who might have sensed the interviewer's intention of feeling for hints of animacy attribution – stressed that for them, the robot was "a pure tool" (i.e. "nothing but a tool"; R7.4-00:06:38-1), that "it stay[ed] a tool" (R2.3-00:02:47-1).

Interestingly, it is exactly this function as an important research tool that also appears to trigger quite a different perspective on the robot. R1 described the robot as a client. They explained that the robot is "the first user of what [the roboticist is] thinking" (R1.2-00:01:08-6): "I work on ... algorithmic functionalities [... which ...] enhance the robot's capabilities, so ... I'm working for the robot" (R1.2-00:02:44-1). This relationship was perceived as reciprocal: not only is the roboticist working for the robot, the robot is also working for the roboticist. When serving as a test platform for new control algorithms the robot acts as a feedback-giver. When a robot is tested with a new version of a control algorithm, its performance is observed and behavioral data recorded. For example, a log of the executed program can be saved, or the robot's behavior can be recorded on video. This documentation of the robot's performance is then used to improve the control algorithm. In order to be able to progress in their work, roboticists therefore depend on the robot's behavioral feedback – roboticist and robot working together in a reciprocal relationship. As R1

2 Nao is the name of the robot model.

explained: “If I work for the robot to enhance its capabilities, then the robot uses these capabilities to help me” (R1.2-00:03:18-6).

Karin Knorr-Cetina (1997) diagnosed a similar reciprocity in her research on sociality with objects. She observed a “mutual providing of self and object”, for example in situations where “a scientist tries ‘to make sense’ of the signs given off by an object to determine what is further lacking, and what she should therefore be wanting to do next” (ibid., p. 23). Knorr-Cetina goes so far as to argue that this kind of human-object mutuality and solidarity, “through the interweaving of wants and lacks specifies a kind of backbone of reciprocity for an object-centered sociality” (ibid., p. 22).

This reciprocity is also reflected in how the interviewed roboticists perceived the distribution of agency between themselves and the robot. They described the robot’s performance as a reflection of both their agency and the robot’s own agency, both perspectives contributing to the robot being perceived as an animate entity. On the one hand, several interviewees explained that their robot’s behavior was always a representation, or even embodiment, of their and their (human) colleagues’ work. It was clear that “there is a lot of the programmer … in the robot” (R3.2-00:03:07-6). Here, the driving force for the robot’s actions was perceived to be the roboticist responsible for the program the robot was executing: “You do always know that [the robot’s behavior] is caused by me, and it’s not like the robot made a rational decision independently”[3] (R7.5-00:02:22-9).

A similar perceived extension of the self, or of other humans, into a robot was observed in earlier research. Thomas Fincannon and colleagues (2004) described how members of search and rescue teams perceived their non-humanoid remote-controlled robot both as an embodiment of the other human team members controlling the robot from a distance and as a team member in itself. This perception was even reflected in their behavior. They treated the robot with similar social rules as other humans, for example by keeping a certain spatial distance, holding eye contact with the robots’ camera, and preferring when the robot faced in the appropriate direction.

In the present study, despite the perceived presence of the roboticist in the robot, robots were also perceived to be their own source of agency. This was especially the case for situations in which a robot behaved in an unexpected way, for example, when it malfunctioned. R1 noted that “[a malfunction] is

3 Translated from German by the author.

[the] fault of the machine, but also of the humans" (R1.3-00:01:08-5). They explained further:

> "When the robot deals with some situations ... you are all happy, 'Ohh, he can make, she can make it', 'it worked'. Okay, from one side, [it] can also be kind of [a] cheer for our own work, if it works. But also it is kind of part on the robot [laughs]. So, some of the credit goes to the robot [laughs]." (R1.4-00:03:03-0)

Here, we can observe a perceived distribution of agency between the roboticist and the robot (cf. Rammert & Schulz-Schaeffer, 2002b). One could even interpret it as a distribution between two equally important actants, in the sense of Actor-Network Theory's generalized symmetry (ANT; e.g. Latour, 2005; cf. Chapter 2, Section 2.3). In any case, the quote above illustrates nicely how this distribution is often enacted: By a constant switching of attributions on a discursive level. This switching was also observable with R2, who changed the subject of one sentence between themselves and the robot several times: "The robot can do, ... you can do dozens and dozens of things, it can do" (R2.3-00:08:38-1).

The unique collaborative relationship between roboticists and their robots was, however, most obvious when several of the interviewed roboticists explicitly called the robot a research companion, or even a member of their research team:

> "It's kind of part of [the] family." (R1.3-00:00:03-0)
> "It is a part of the team, it's a fact!" (R1)[4]
> "It's more of a team member." (R4.1-00:07:14-9)
> "They are team members in some sense." (R8.4-00:05:08-8)
> "... some kind of research companion." (R2.3-00:02:47-1)

Interestingly, the robots' role as team members was not obviously reflected in their given names. Only few robots were referred to by a human name. Some were named with an acronym, often involving a pun. Some were not given an individual name at all, but instead were referred to by the brand or model name, or an identifying inventory number. This was in spite of the robots being perceived not as any other team member, but as one making a significant contribution to the R&D process. After all, without the robot's

4 No audio recording timestamp available as the statement was made by R1 after the end of the official interview.

constructive feedback, the roboticist would not be able to continue improving their control algorithms. R3 explained:

> "It's really important to have the robot as a member of the team because it's actually a really valuable asset. ... In robotics, there is something that happens that if you don't have [the robot] you can't work." (R3.5-00:07:00-7)

This dependency on the robot coworker and its influence on the robot's perceived value as a team member was similarly observed by Julie Carpenter (2013). In a study on military personnel employing robots for explosive ordnance disposal (EOD) Carpenter observed that the (non-humanoid and non-autonomous, i.e. remote controlled) robots were sometimes perceived as valuable team members. Also in this specific life-and-death field of application, there was a strong dependency on the robot's performance. Deploying a robot to remove explosive ordnance, and to potentially be destroyed in an explosion, palpably saves the lives of those soldiers who without the robot would have to approach the ordnance themselves (Sandry, 2015b). This lifesaver role of EOD robots is such a strong narrative that even the US Department of Defense's press department regularly uses it in the context of their public relations efforts (Roderick, 2010; cf. Chapter 4, Sections 4.3 and 4.4).

3.5. Testing in the Real World: The Unpredictable Robot

In the context of robotics R&D, a robot's behavior during its crucial job as feedback-giver is one of the strongest perceptual triggers of animacy attribution. After all, the most interesting feedback for a roboticist is when the robot does not what it is supposed to do. This is inevitable, when a system as complex as a robot is let loose in an environment as complex as the real world. Transferring work from simulations to a physical robot is rarely a smooth process, as R3 explained:

> "From the point of view of the experimentation you can test with images, or with data, recorded data. But then, when you go to the actual robot, a lot of things happen. ... Things are moving, changing. ... We live in an uncertain world. And only when you get to that stage [of experimentation] you realize [that]." (R3.1-00:08:59-0)

> "Working with the platform is ... not as comfortable as working with your computer. ... There is always something that doesn't work." (R3.1-00:04:14-6)

The frustrating process of dealing with unexpected robot behavior, tweaking and tinkering until the robot does what it is supposed to do, is a central aspect of this particular phase of robotics R&D. Robot platforms can be incredibly complex pieces of hard- and software. This is especially the case in projects where a heterogeneous team of roboticists works together on a shared customized, or even custom-built, platform. Some interviewees explained that their knowledge of the robot's complexity alone, without even observing its behavior, was enough for them to attribute animacy in the robot:

> "[The robot is] something that has different modes, different systems, that is moving around ... Something that has some process going on is something lively." (R1.4-00:02:03-5)

> "As the system is so complex it also has some kind of life of its own." (R4.1-00:09:00-0)

Sometimes it is actually not a robot's complexity, but its "dumbness" that causes unexpected and seemingly intentional behavior. R3 described a robot that was supposed to ignore certain light sources in the environment. When this functionality failed, the robot oriented its movements towards lights in the room. The roboticists then explained to visitors that "[it] likes the lights" (R3.2-00:06:15-5). The seemingly intentional behavior of the robot was actually the result of a malfunction. For the interviewed roboticists exactly these "dumb" behaviors acted as the strongest triggers for animacy attribution, by making the robot appear willful. R3 observed: "When [the robot] gets stuck all the time in a particular place ... it kind of gives [the robot] a bit of personality" (R8.3-00:06:39-4).

Crucially, a roboticist might have expert knowledge only about one particular component of a robot platform. The technical intricacies of other components – those, which other team members are responsible for – might be partly, or even completely, closed off to them. In practice, this means that when a system as complex as a robot is deployed in the real world – no matter how closely controlled the experimental setting is – its performance can be quite unpredictable. Noisy data transmission, ambient temperature, floor texture, air movement, power fluctuations, spilled coffee, crumbs lodged in a crease, vibrations from a train passing by outside: countless environmental factors can disrupt the proper functioning of the robot, or at least lead to situations which would have never occurred in a simulation. These challenges apply especially to application scenarios that are typical for "New Robotics", where robots are not confined to structured environments but expected to act and navigate in very complex, frequently changing surroundings (cf. Chapter 1, Section 1.1). R4 described how their robot struggled with the summer heat:

> "It happens frequently that some kind of errors occur which you can't explain. ... Sometimes we have problems with heat dissipation. In the summer, when it's quite hot, the computer fans will go faster and faster, and at some point he just doesn't want anymore." (R4.2-00:00:55-3)

The more realistic the experimental setting the more can go wrong. Especially challenging scenarios involve humans – their performance is notoriously uncontrollable. Consequently, working with a physical robot in an experimental setting almost inevitably leads to situations in which the robot does not behave according to any preexisting plan, but simply "does what it wants". Practical roboticist work therefore involves dealing with countless breakdowns and interruptions, as well as hours of troubleshooting and debugging. Even the most experienced roboticists will have to face situations in which they do not know why the robot just did what it did. When working with autonomous robots, a roboticist might have no information at all about what is happening "inside" the robot. It becomes a black box – meaning that what exactly the control algorithm is "doing" at a certain moment can only be inferred from the robots observable behavior.[5] It is these situations in which attributions of animacy are most common. R4 explained how they observed themselves having two competing explanations for a robot's unexpected behavior:

> "At some point [the robot system] gets arbitrarily complex and what you're looking at in the end is the [robot's] behavior. And if you're immersed in the subject matter then maybe you can explain the [robot's] behavior. ... If you aren't immersed in the subject matter, then you only see the behavior. Which you then can either explain in a more abstract way: 'Well, the intention was for him to drive from A to B' or something. 'And for some reason this doesn't work that well right now'. Or you see it as 'Well, he has some kind of task and

5 Cf. the concept of the "nontrivial machine" (von Foerster, 1993).

> for that he has to drive from A to B, and he hasn't done this task that well. He must have a bad day." (R4.4-00:02:10-0)

R4's description of their two very different interpretations of the robot's behavior can serve as an example for the final crucial observation of this chapter: For the roboticists interviewed for this study, attributions of animacy and inanimacy were reflections of the changing perspectives they routinely take on their work.

3.6. Switching Perspectives: In/Animacy Attributions as Constructive Practice

Roboticists in academia do not only work in closed-off laboratories, they do not only talk about their work among like-minded peers with similar expertise. Just as in other disciplines, their work is embedded in the complex environment, practices, and discourses of their field – and sometimes even breaches the border to the realm of entrepreneurship. They present their work in discussions, presentations and publications, to academics of other (sub-)fields, to reviewers, to funding agencies, and to the lay public – be it indirectly through journalists, or directly in science communication contexts (which we will explore further in Chapter 4). Researchers balance the – often contradicting – demands and needs of their academic peers, their disciplines, their funding organizations, potential investors and customers, and the wider public (Möllers, 2015, p. 143; cf. Jasanoff, 2001). In practice, they "shift... from one social world to the other by mobilizing different cultural registers" (Möllers, 2016, p. 19). For the roboticists interviewed for this study, this switching of worlds was also mirrored in their switching of perspectives and of attributions of in/animacy to robots.

A roboticist's perspective on their robot can also change with the stage of the research process. A robot can at first be a tool for the development of a software, which is later used on the same robot, which is then perceived as a demonstrator or a completed creation. R2 described their constant change of perspective:

> "I work with the robot. So, not on the robot. ... Eventually, when the work is finished after all the years [laughs] then I actually worked on a robot control system. Only to get there, now, I have to work with the robot." (R2.3-00:01:19-1)

R4 explained that their attributions of animacy differed strongly between the robot they worked on (or with) in the laboratory, and robots they encountered in other professional situations like a conference or trade fair, or in their private life – for example when using a lawnmower robot. They even switched perspectives on the lawnmower robot, viewing it either from what they called a "psychological" perspective, where the robot's behavior would be attributed to some kind of personality ("has a bad day", R4.4-00:01:49-0), or from a "technical" perspective, where the behavior would be explained via the robot's control algorithm and its interaction with the physical environment.

Specific discursive practices were likewise dependent on the social context. In some cases, animacy attribution was reflected in roboticists using gendered pronouns ("he", "she") for their robot. However, some apparently perceived their own behavior as unprofessional. During the interview, they interrupted themselves with self-conscious laughter when talking about gendering the robot. Some automatically used a gendered pronoun when talking about their robot, but quickly corrected themselves, repeating what they said with the neutral pronoun "it". R3 mentioned that they had to take care not to use gendered pronouns in scientific publications – presumably, because that would be considered unprofessional:

> "... something that you put in a robot and he start[s] trying... 'It'. Sorry, I always say 'he' [laughs]. ... I have to change all the text I write. [laughs] To write 'it'." (R3.3-00:07:20-1)

The roboticists' constant switching of attributions was most obvious on a linguistic level, their wording frequently reflecting the apparent contradiction of animacy and inanimacy. For example, R4 observed that for them the robot was three things at once: "team member, tool, platform to test things" (R4.1-00:09:00-0).

Phrases like "but still...", "but somehow", "but also...", "or actually...", "but maybe" were uttered by almost every interviewee in this specific context.

> "You are aware that it is a technical object. But still, when the robot deals with some situations and some things you are all happy, 'Oohh, he can make, she can make it', 'it worked'." (R1.4-00:03:03-0)

> "Of course it's a machine. But when it gets stuck all the time in a particular place. I don't know, it kind of gives them a bit of personality or something." (R8.3-00:06:13-3)

> "You never think that the robot is doing something intelligent somehow, because there is a lot of the programmer still in the robot. But somehow you ... sometimes you just think it's doing something you never expected it to do [laughs]." (R3.2-00:03:07-6)

In some cases, the multiplicity of perspectives took on almost absurd dimensions. For example, R2 stated that their robot "is [a] robot companion. Some kind of research companion" (R2.3-00:02:47-1), and then immediately changed their mind: "Actually that goes too far, 'companion', because I would say it is a research tool. So, it is a tool, it stays a tool for me" (ibid.). R8 took the other direction, first stating that they "don't think anyone here feels so much attached to robots ... that they say that they see them as pets or colleagues" (R8.4-00:05:08-8), and then immediately adding that "of course, they are team members in some sense" (ibid.).

While some denied attributing any animacy to their robots, others openly embraced it, and understood it as a natural aspect of their work life. For example, while R2 stressed that the robot was "a pure technical object" (R2.2-00:00:50-0) and "nothing else" (R2.2-00:01:54-0), R5 observed that "[this] does happen when you deal with a robot for such a long time. You do humanize the robot" (R5.3-00:01:37-0). R1 outright stated that their robot "is a part of the team, it's a fact!".[6]

Julie Carpenter (2013) similarly found that the members of military bomb disposal squads she interviewed had contradictory emotions and reaction to their robots, fluctuating between playful affection and awareness of the robots' inanimacy. They described their robots (and not other machines) with anthropomorphic and zoomorphic terms, at first openly admitting to anthropomorphizing their robots, but then immediately downplaying it, explaining it as a joke. Also here, the constant switching of perspectives seemed to be a constructive way of dealing with the complex, sometimes contradicting, challenges of the professional work environment: The soldiers depended on the robot because it spared them from having to approach explosive ordnance – giving it the status of a valuable team members –, but also had to be able to send the robot to its possible "death" without hesitation.

6 No audio recording timestamp available as the statement was made by R1 after the end of the official interview.

Similarly, for our roboticists, while the superficial effect of switching perspectives and attributions is that of an ambiguous animacy, it is not merely a reactive practice. It is not only a way of dealing with the peculiarities of the technological artifacts they are exposed to on a regular basis within the context of their profession. It also is a constructive practice, in the sense that being able to take on different perspectives on the robot enables roboticists to perform adequately in different contexts of their profession. It reflects roboticist's parallel commitments to research, development, science communication, and entrepreneurship. On the one hand, it enables roboticists to see their robot as a tool for, or product of their work, and to be able to and to keep the emotional distance this requires. On the other hand, it enables them to acknowledge the robot's autonomy, the importance of taking its sometimes seemingly erratic behavior serious, and to use it constructively as feedback for the further progress of their work. Within a heterogeneous R&D team, playful references to the robot platform as an animate being reflect its agentic role as a feedback-giver and constructively supports the team's communication and collaboration.

3.7. Summary

Most research on human-robot interactions focuses on short-term encounters of lay users with robots in more or less artificial experimental settings. So far, little attention has been directed towards the context of robotics laboratories, which can provide a unique opportunity to study long-term interactions of humans with robot technology in a professional context. The present chapter explored this context with a small interview study, taking special focus on roboticists' attributions of animacy to their robots.

Despite their expert knowledge of robot technology, and their presumed professional distance, most of the interviewed roboticists regularly attributed animacy to the robots their worked with. They routinely took multiple, sometimes even seemingly contradicting, perspectives on their robots, constantly switching the level of animacy they attribute to them. These attributions were shown to be partly a reaction to the robot's design and behavior itself, and partly a reflection of the roboticist's unique and challenging work environment.

Roboticist teams' focus of attention is directed towards the robot by several interconnected factors: A strong political demand for applicable and marketable results, as well as demonstrable platforms, and the resulting challenge of experimental work; the typically disciplinary heterogeneous team structure, for which the robot platform serves as a boundary object by providing a shared focus of attention; and finally, the lack of standardized hard- and software infrastructure, and the resulting necessity of customization work.

Robots were shown to have a crucial role as feedback-givers in robotics R&D. The dependence on robots' contribution to the R&D process left the impression of a distribution of agency between the roboticist and the robot. Moreover, in the highly application-driven environment of robotics R&D, testing robotic platforms' performance not only in simulations but also in real life is a central aspect of roboticists' professional work. Through the friction caused by the interaction of complex robot technology and complex physical environment, roboticists are routinely confronted with unexpected, unpredictable robot behaviors. These can serve as a strong perceptive trigger and cause roboticists to attribute a self-will or personality to the robot. These attributions expressed themselves in the form of gendered language, but also in the form of references to the robot platform as a biological-appearing body, with body parts and even physiological processes. The most frequently observed practice was, however, that of regarding the robot platform as a valuable (if often uncooperative) member of the R&D team.

Crucially, we were able to observe that these attributions of animacy are not one-sided or enacted in a static or forceful way. Instead, the enactment of animacy attributions is highly situation specific, roboticists constantly switching between discursive and practical representations of the robot as an inanimate tool and the robot as an animate team member. This switching is not only a playful way of dealing with the unique characteristics of robot technology, but also a reflection of roboticists' professional practice. Attributions of animacy thus are an integral and constructive practice of robotics R&D, in that they allow roboticists to adapt flexibly to the parallel commitments and challenges of research, development, science communication, and entrepreneurship.

In other professional contexts, however – such as written research papers – attributions of animacy would be considered unprofessional, necessitating a different perspective: that of the robot as an inanimate machine. Then again, when promoting their R&D work outside of the immediate circle of academic peers it can be beneficial for roboticists to stage a narrative of the robot as an animate being. The next chapter will explore this context and show that even more explicit attributions of animacy to robots can be observed in demonstrations, science communication, and marketing.

4. Showing Off Robots: In/Animacy Attributions in Robotics Demonstrations, Science Communication, and Marketing

4.1. Demo or Die: Outreach, Engagement, and Accountability

Just as other scientists and engineers, roboticists routinely present their work to academic peers and sponsors, as well as to potential customers and the lay public – and they are expected to do so in an increasingly professionalized manner. They stage live and video demonstrations, are involved in science communication efforts, and those who are (also) entrepreneurs have to engage in public relations and marketing as well.

> "Researchers always have something to sell. ... Those working in academia are looking for talk invitations, citations, promotions. ... Those working in a large company will want to create interest in some product." (Togelius, 2017).

The present chapter will explore this context and show that, also in this professional environment, attributions of animacy to robots are an everyday practice – and not only as a reaction to robots' unique characteristics, but in fact as a constructive aspect of robotics demonstrations, science communication, and marketing.

Across all disciplines, scientists and engineers are required by overarching science policies and individual funding institutions to present and promote their research efforts – not only within their immediate disciplinary communities, but increasingly also to the general public. These expectations are part of a pervasive "rhetoric of 'outreach' [and] 'engagement'" (Weingart, 2019), which is reflected in broad efforts like the PUSH memorandum – a German initiative calling for a stronger engagement for the communication of scientific results to the general public (Wissenschaft im Dialog, 1999). Increasingly, specific requirements for public engagement are also inscribed in grant contracts with funding organizations. For example, projects funded by the European Commission under the Horizon 2020 scheme are required to "promote [themselves] and [their] results, by providing targeted information to multiple audiences (including the media and the public), in a strategic and effective manner" (European Commission, 2017, p. 71).

Consequently, demonstration and science communication practices have become increasingly professionalized and are often driven by institutions' longer-term strategies – not unlike corporate communication efforts (Trescher, 2010). Moreover, instead of just being able to delegate communication tasks to their institutions' press department, scientists are often encouraged, or even pressured, to act "as their own sender" (Trescher, 2010, p. 27; cf. Leopoldina, Acatech, & Akademienunion, 2014, 2017). One consequence of this is a growing relevance of social media for science communication, as it offers a relatively easy way for researchers to draw attention to their work and to connect with an interested audience (Leopoldina et al., 2017; cf. Könneker, 2019).

Roboticists, too, are under immense pressure to legitimize their work, in order to justify past and future funding, to ensure public support, and to meet their "democratic obligation of accountability" (Weingart, 2019). Consequently, "robotics researchers are investing considerable time and effort in 'engaging' publics" (Wilkinson et al., 2011, p. 367). Not only do they have to comply with funding institutions' requirements for dissemination activities. In robotics, even some academic journals require that each article is supplemented with a demonstration video. Also live demonstrations for sponsors, potential customers, and the general public are a regular aspect of roboticists' professional lives.

Robots on Social Media

Many individual roboticists, robotics institutions, and businesses present their work on social media. A practice rather specific to the robotics field is that of running a social media account not (only) for a whole institution, research group, or brand, but for just one specific piece of technology: for a certain robot model, or even for a singular robotic individual. NASA[1], for instance, has been running several Twitter accounts for over a decade. This does

1 National Aeronautics and Space Administration (USA).

not only include accounts for whole institutions, such as the Jet Propulsion Laboratory (JPL), but also accounts for individual pieces of technology. At the present time, there are at least six NASA spacecraft and three planetary rovers with their own dedicated Twitter accounts. These accounts give regular updates on the craft's activities and refer to information and news from the wider space flight community.

Crucially, the tweets posted by several of these accounts are written from the perspective of the spacecraft and rovers themselves. In 2008, JPL's social media team first started letting the Phoenix Mars Lander[2] "tweet" in the first person perspective, and discovered that these tweets gained more reactions from followers: "The first person robot is what breaks the ice and gets people feeling like there's a conversation going on" (Li, 2014). Since then, more NASA spacecraft and rovers, as well as ones from ESA[3] and ISRO[4], have joined the club, tweeting – with varying frequency, and with varying payoff – in the first person perspective.

This unique practice of making robots speak for themselves is highly instructive for the way narratives of robot animacy are utilized in the science communication and marketing context. They will therefore play a central part in the present chapter. While the described "space robots" (spacecraft and planetary/asteroid landers and rovers)[5] are by far the most popular examples, there are also many other types of robots with dedicated social media accounts. Some document the "lives" of bespoke humanoid robots serving as a kind of ambassador for their research groups. Unlike the spacecraft and rovers far away from earth, these robots are usually also regularly presented in live and video demonstrations.

Robot Demonstrations

Demonstrations, just as the communication efforts described above, are an increasingly mandatory and professionalized aspect of robotics research

2 https://www.nasa.gov/mission_pages/phoenix/main/index.html (accessed 2019-12-21).

3 European Space Agency.

4 Indian Space Research Organization.

5 These planetary and asteroid rovers and spacecraft are so-called mixed initiative or shared autonomy systems. While they receive high-level instructions from human operators, more low-level behavior, such as obstacle avoidance, is controlled autonomously by the rover/spacecraft (Richards & Smart, 2013).

and development. In this, they complement the infamous academic mantra "publish or perish" with "demo or die" – an idea attributed to MIT Media Lab founder Nicholas Negroponte, who demanded that researchers and engineers focus on producing artifacts (instead of only publications), which then could be demonstrated to the lab's corporate sponsors (DuVergne Smith, 2014; Markoff, 1996).

Robotics demonstrations typically consist of a robot performing specific tasks, often visualizing a use case relevant to the intended audience. Usually, the robot's task (such as "grip object and move to new location") is embedded in a short narrative ("robot serves drink to person") and a scenario fitting the application goal of the overall project ("at-home care of an elderly person"). In commercial contexts, the goal is to pitch the robot to potential investors or customers. In an academic context, demonstrations can have several overlapping functions and target audiences: They can be a routine part of the publication process, the audience being peers in the academic robotics community. They can also be targeted towards funding agencies and industrial sponsors (Rosental, 2005). A demonstration might be used to visualize current or anticipated robot abilities in a grant proposal, or as part of the reporting process of an ongoing project. Last, but not least, robot demonstrations can be geared towards the general public. On the one hand, successful demonstrations are used to legitimize past funding – "proving" that a research project was worthy of the funds invested in it. On the other hand, showing what robots can do can also be meant to "calibrate the public" and the robotics community itself to the current state of the art (Pratt, cited in Belfiore, 2014).

This shows how in robotics, science communication in the traditional sense (i.e. practices meant to communicate research results to the lay public) often overlaps with practices of presenting results to peers and sponsors in the scientific community, and with practices of marketing products to potential customers. It is not uncommon for roboticists to launch start-up companies, selling technology developed previously in an academic context. In these cases, demonstrations and science communication activities double as marketing activities. This is also observable in the combination of demonstrations and dedicated social media accounts, which are common in both academic and commercial robotics. Using the terms employed in the context of the EU Horizon 2020 program: in robotics, it is difficult to draw a clear line between the communication, the dissemination, and the exploitation of research results (European Commission, n.d.). This blurring of boundaries makes it necessary to take the different areas of practice into account together. The present chapter will show that they have the same functions (such as providing apparent proof of a robot's functionality), that they employ the same techniques (such as references to popular fictional narratives), and that in doing so they all end up "staging" robots' animacy.

4.2. Approach

Cases and Method

Many commercial robot developers not only present their robots in live and video demonstration, they also run social media accounts documenting their robots' "lives". Just as in the academic context, one can encounter both bespoke and small-series humanoid platforms, such as Hanson Robotics' Sophia and Boston Dynamics' Atlas. There are smaller humanoid platforms, such as Softbank's Nao and Pepper – which are by now robust, affordable, and easy enough to use to be marketed not only as research platforms but as programmable education, entertainment, and service robots. Finally, there are household robots such as iRobot's Roomba – featuring technology that is already decades old and by now cheap and robust enough to allow mass production and success on the mass consumer market. Together with the space robots introduced in Section 4.1, these robots make up part of the diverse sample on which this chapter's observations are based. A complete list of all cases is available in the Appendix.

These cases cover a wide spectrum of activity (ranging from one demonstration video every few months, to several social media posts every day), success (from barely any engagement, to millions of followers and video views), professionalism (from a lone researcher dabbling in social media, to a team of trained marketing and video production staff), and interactivity (from quietly posting a video and leaving it be, to complex scripted interactions with other communication teams across multiple platforms and lively engagement with social media followers).

This sample, and the analysis based on it, do not strive to give a comprehensive image of the whole landscape of robotics science communication, demonstration, and marketing. Rather, the cases examined in this chapter were chosen for their potential to illustrate this book's specific point of interest, that is, the attribution of animacy to robots. This is why, for example, industrial robots do not feature in the sample. For the same reason, the quotes and examples presented in the following sections focus more on social media activities, and less on other science communication and marketing activities such as press releases, articles and interviews in the popular press, trade fair visits, open lab days or science slams.

The websites, as well as social media, marketing and demonstration activities and media reports on each case were tracked for a time period including the year 2017 and the first half of 2018. Especially instructive events and documents from before and after this period of time were included in the corpus as well. A special focus was set on the Twitter accounts connected to each case, with all tweets from the specified time period documented and analyzed individually. As in the previous chapter, this corpus of material was systematically analyzed following a qualitative content-analytic approach (Mayring, 2010). Again, analytical categories were developed inductively and iteratively from the material, the central criterion being instances of animacy attribution to robots in the wider sense (including attributions of physiology, sensory experience, cognitive processes, intentionality, sociality, personality, emotion), as well as hints to practices of staging robot agency and animacy (e.g. in the form of a purposeful backgrounding of remote controlling of robot activity).

Here, too, the goal of this process was not to measure or quantify the "amount" of in/animacy attribution, but rather to document the qualitative range of attribution practices, in order to then identify the context, strategic function, and consequences of in/animacy attribution practices in each specific instance.

Chapter Structure

With the help of the cases introduced above, this chapter will explore how robotics demonstrations, science communication, and marketing practices skillfully utilize attributions of animacy and inanimacy to robots for their respective goals.

First, the chapter will show how a staging of robot autonomy and animacy, together with a backgrounding of human agency, are used to provide proof of robots' functionality (Section 4.3). Second, we will see that robots are embedded in scenarios of trouble-free use and narratives of desired futures in order to demonstrate their relevance and applicability (Section 4.4). Third, the chapter will show that robots are made tangible and exciting for the audience by embedding them in engaging narratives, featuring them as animate single entity personas capable of social interaction (Section 4.5). We will also see that, in all these contexts, practices of animacy attribution are not performed consistently, but are instead one aspect of a constant switching of narrative perspectives on the robot as an animate (appearing) autonomous being or a human-controlled machine (Section 4.6). Finally, the chapter will show that these practices are sometimes criticized for causing misconceptions and bias (Section 4.7).

4.3. Narratives of Agency: Proof of Functionality

In most academic disciplines, the default path of presenting research and engineering work to peers and the public is through the publication of written articles. A description of research methods, results, and conclusions, presented in a manner sufficiently convincing to reviewers, is understood to serve as proof for the reported findings and successes. In technology development, it is common practice to add another level of proof: In order to show that, for example, a robot is functioning as promised in a research article, funding application, or PR brochure, demonstrations are performed "to show the feasibility of a technological approach, the value or even correctness of a specific technological approach, ... or the proper running of a prototype or product" (Rosental, 2005, p. 346). In robotics, providing a demonstration video is sometimes even a prerequisite for the publication of a peer-reviewed article.

Strictly speaking, a demonstration is only able to prove a robot's functioning in the moment the demonstration is performed. In practice, demonstrations also are understood to "imply that what might have only worked once will work anytime, anywhere and without the implicated networks of human and nonhuman actants" (Both, 2015, p. 1; cf. Suchman, 2007, p. 148). This has two major consequences: Demonstrations routinely are carefully scripted and rehearsed performances and any "unseemly" human intervention is usually backgrounded, or even concealed.

Especially in the context of robotics, autonomous and robust functioning is a central goal. However, technological progress in robotics can appear frustratingly slow to the uninitiated observer. Roboticists are usually aware that somebody outside of their specific field cannot appreciate the technological significance of a seemingly small and unimpressive improvement in robot performance. Thus, as a robot's performance in a demonstration is understood to stand for its performance in the future, it is crucial that everything proceeds perfectly as planned. As a consequence, robotics demonstrations often employ a variety of staging techniques.

A robot might not (yet) be able to interact smoothly with a user via natural language interface, as promised in a project's funding pitch. Hence, for a live demonstration, test users might be briefed to use specific verbal commands, or even be trained to use a certain tone of voice that is easily understandable to the robot (Lipp, 2017, p. 122). For big commercial demonstrations, such interactions are often even scripted word by word (cf. Sharkey, 2018).

Demonstration videos sometimes use time lapses, showing a robot's movements sped up considerably. A video presenting an autonomous towel-folding robot received considerable attention at the time (UC Berkeley Robot Learning Lab, 2010). The video is sped up by the factor 50, veiling the fact that the robot takes 20 minutes to fold one towel. The intention is to make the video shorter and more interesting, but it can also be a trick to conceal a robot's slowness and to make it appear more dynamic and agentic. Usually (but not always) the applied speed factor is disclaimed somewhere in the video. Nonetheless, a time-lapse video makes it difficult for the audience to get a realistic impression of the robot's actual performance speed.

Moreover, demonstration videos are usually edited to include only successful performance trials. One of the roboticists interviewed for Chapter 3 of this book explained:

> "If it works perfectly one time, and then you see a video of that [then you] think 'Ok, works'. ... But maybe you even know, but you would never write that in the paper, that it wouldn't be applicable in reality".[6] (R7.6-00:03:54-2)

A demonstration video showing a robot hand successfully "solving" a Rubik's cube, which received considerable attention by the press and on social media, was called out for concealing that the robot apparently only was successful in 20% of the trials (Marcus, 2019a).

Practices like these are aimed at controlling what Catelijne Coopmans (2011) terms the "face value" of a technology. This apparent value (in contrast to the actual value) "focuses attention on the visible surface, on the 'face' that gets presented or shown" (ibid., p. 158) and does not necessarily match reality. There are several facets to the reality that is so carefully shrouded. On the one hand, it is the reality of the robot's performance – for example its slowness or unreliability. On the other hand, it is the reality of human involvement in

6 Translated from German by the author.

its performance. This does not only apply to technology demonstrations, but also to other academic contexts, in which the messy reality of research work is carefully kept away from an audience. Stephen Hilgartner (2000, p. 19) points out "the differences between formal scientific texts and the activities required to produce them ...: scientists tinker in the privacy of the laboratory until they are ready to 'go public' with neatly packaged results".

The staging of technologies for the sake of making an impression on certain audiences is not a modern phenomenon. David Gooding and Frank James (1985) described how nineteenth-century scientist Michael Faraday strived to make phenomena demonstrable and self-evident by artfully backgrounding any human involvement in the phenomena shown in a demonstration (cf. Golinski, 1998, p. 94). Also Steven Shapin and Simon Shaffer (1985), in their work on the air-pump experiments conducted by seventeenth-century chemist Robert Boyle, described how public demonstrations of the experiments were carefully scripted and staged, hiding the fact that the shown "effects of nature" were actually controlled by human actors.

These efforts to obscure the involvement of human agents in a technological performance are sometimes compared to techniques employed by stage magicians. Similarly to magic shows, demonstrations reach their desired effect though "the combination of simulation and dissimulation: creating an effect known by all to be contrived, while simultaneously erasing signs of its contrivance in machinery and method" (Alač, Movellan, & Tanaka, 2011, p. 336). Other authors make the connection from technology and science demonstrations to theater studies as well. Norma Möllers (2016) observed that scientists, in order to stage their work as applicable, performed a "technoscientific drama" when communicating with their funding institution. In this, Möllers draws on Goffman's (1959) concept of self-presentation, which distinguished between "front stage" behaviors, which are meant to be visible to the audience, and "back stage" behaviors, which are only shown when no audience is present. In the case of technology demonstrations, a "back stage" action could be, for example, the hiding of a mess of cables and unsightly equipment under a tablecloth, or the hectic commotion of assistants behind a partition, making last-minute corrections in the program code of the robot demonstrator. Bruno Latour drew on theater metaphors as well, coining the term "theater of proof"[7] for situations in which scientists "'force' [the audience] to 'share' [their] point of view" (Latour, 1993, p. 86). Andreas Bischof and Göde Both

7 Depending on the source sometimes called "theater of truth".

(2015) called robotics demonstration videos a "cousin" of Latour's theater of proof – both employing powerful orchestrations in order to make the existence of a certain phenomenon obvious and self-evident. Bischof (2015, p. 286) points out that robotics demonstrations are social events with unique rules. He observed that, rather than worrying about a robot's epistemic features, roboticists often focus on its "stageability"[8]. Both (2015) introduced the term "Youtubization" to describe practices of embedding a robot demonstrator in dedicated choreographies and narratives. He observed that, in some cases, these were staged to such an extent that they "d[id] not … conflate with the project's overall objectives and work practices" (ibid, p. 3).

Probably the most frequently staged aspect of a robot, both in academic and commercial contexts, is its autonomy – its independence of human control. Recent examples are small mobile robots used in the United States to deliver fast food to customers' doorsteps. While appearing to move autonomously, they are actually remote controlled by minimum-wage workers in Colombia (Said, 2019). In the wider context of commercially used artificial intelligence (or rather, pseudo-AI), practices of employing humans as "mechanical turks"[9] or "ghost workers" (M. L. Gray & Suri, 2019) have reportedly lead to absurd situations: The "automated" office assistance offered by the company X.ai, for example, is actually performed by human employees – which are required by the company to interact in messages to users in a "robotic" way in order to leave the impression of interacting with AI (Lobe, 2019).

Demonstrations, too, often hide that a human is remote controlling certain robot functions, or that the robot actually only manages to complete its task in one out of dozens of trials. In live demonstrations, a human controller might be hidden off stage; in a demonstration video, they might be kept out of the camera's shot. Lucy Suchman observed such an "erasure of human labors" (2007, p. 238) in her studies of robotics laboratories. She noted that a robot in its "'backstage' environment provided an opportunity to see … the extended network of human labors and affiliated technologies that afford[ed] its agency" (ibid., p. 260). Suchman interpreted this as a "lesson… requir[ing] that we reframe [the social robot] from an unreliable autonomous robot, to a

8 German "Inszenierbarkeit" (translated by the author).

9 The Mechanical Turk was a chess-playing machine from the late eighteenth century. It was presented as being able to play chess autonomously, but in fact was controlled by a human hidden inside the machine (e.g. Standage, 2002).

collaborative achievement made possible through very particular, reiteratively developed and refined performances" (ibid.). Frequently, these performances feature "extreme and spectacular circumstances ... in order to impress the audience and in order to produce witnesses of the achievements shown on stage" (Rosental, 2005, p. 346). One of the roboticists interviewed for Chapter 3 of this book explained that "you will see that a lot in robotics. Like, robots juggling objects or something. Just to impress laypeople" (R8.1-00:08:57-1).

Especially interesting examples are the impressive demonstration videos of Boston Dynamics' biped humanoid and quadruped robots (Boston Dynamics, n.d.). The company is famously secretive about the technical details of its work, but regularly releases spectacular videos showing off the robots' newest abilities. Whether it is the dog-like Spot robot opening doors or the humanoid Atlas gracefully leaping over obstacles – the videos regularly go viral and gather millions of views (e.g. Boston Dynamics, 2018b, 2018a). Boston Dynamics' video demonstrations are so impressive because the robots appear to have physical abilities surpassing those of most other state-of-the-art robots. Moreover, these videos make the robots appear completely autonomous. Rarely is there any human visible near the robot, the cameraperson seemingly being the only one following it around. In reality, Boston Dynamics' robots are only partially autonomous. Most are remote controlled by a human operator – a fact that is usually carefully concealed in the viral videos. It is in the company's interest that the videos remain vague on technical details, but full of fuel for speculation about the robots' abilities. This "helps to create interest around Boston Dynamics' projects, and their ... secrecy insures that competitors cannot copy their achievements, strikes the public's imagination, and leaves everyone in the dark about the weaknesses of their technology" (Shih, Sinapayen, & Kurenkov, 2019).

An intentional backgrounding of human labor can also be observed in more traditional science communication contexts. Ian Roderick (2010) found that the US Department of Defense frequently embeds its military robots used for explosive ordnance disposal (EOD) in a "life-saver" narrative. The remote controlled robots are presented as if they were able to act autonomously, and the human controllers' involvement is backgrounded: "The [EOD] robot is represented as being able to 'peer around doors', 'carefully adjust its height to survey', and roll 'carefully towards suspicious looking vehicles'" (ibid., p. 239). This "create[s] a sense of automation and agency on the part of the remote-controlled devices that is actually beyond the technology" (ibid., p. 235).

Also in the cases of robots "posting" on social media in a first-person perspective (cf. Section 4.1), human involvement is mostly backgrounded: The question of who (or what) exactly is writing the tweets is usually left unanswered. Presumably, most readers and followers are aware that humans are running the accounts – but these ghostwriters are rarely explicitly credited. Thus, the robots' tweets and posts create a narrative of functionality in a way similar to a remote controlled demonstration. In both cases, the human agents' involvement is not actively denied - but it is carefully pushed to the background.

4.4. Narratives of Desired Futures: Proof of Applicability

Both, robotics demonstrations and "a robot's" first-person social media posts, present a simplified simulation of the present reality. They highlight and narratively stage a robot's abilities, such as robust functioning and a high level of autonomy. At the same time, both can also be a narrative performance of the robot's anticipated and desired future capabilities – making the robot appear closer to what it is supposed to become with further technological progress, and "provide proof ... of the feasibility of the imagined futures" (Both, 2015, p. 1), of the relevance and applicability of the technology in question. In this, demonstrations routinely perform "relevance staging"[10], a presentation of robots in scenarios anticipating the intended or desired use (Knorr-Cetina, 1991, p. 207).

These practices are partly stimulated by the typical project-oriented funding structure of robotics research. Each new grant proposal has to paint anew a desirable vision of a future featuring the yet to be developed robot technology (Bischof, 2017b; Lindemann & Matsuzaki, 2017). Frequently, these visions feature robots that are much more autonomous, agentic, and even human-like than what the present state of technology has to offer. Jane Calvert calls this practice "tailoring of research", meaning "making one's work appear more applied to gain funding and resources" (2006, p. 208; cited in Both, 2015, p. 24). Typically, these staged application scenarios are tailored to the current political and societal discourse. For instance, with demographic change and the aging society being topics of increasing importance, application visions in

10 German: "Relevanzinszenierung".

grant proposals and demonstrations in service robotics often feature scenarios involving elderly care. Search and rescue is another popular application area – as noted in a satirical cartoon from 2019 (see Figure 3).

These scenarios often imagine a present level of functioning which is not yet realizable at the current state of technology, but which is desired or anticipated for the future. A service robot might be able to drive over to a human user and hand them a glass of water, but only under the very specific, staged, scripted, and rehearsed circumstances of the demonstration – such as the wording and tone of the voice command, the lighting situation in the room, the material of the floor, the shape of the glass and the color of the drink (Lipp, 2017). Crucial limitations like these are usually not made explicit in the context of a demonstration. An uninitiated observer thus can get the impression that the robot would function in any realistic home environment, and would be able to interact smoothly with any uninitiated user.

Demonstrations serve the goal of proving that desirable technological futures are attainable. Ben Goertzel, then Chief Scientist of Hanson Robotics, openly discussed that one goal behind the artful staging of their humanoid Sophia robot was to instill in the audience that Artificial General Intelligence[11] was within reach – even though the current state of technology is in fact nowhere near AGI (König, 2019):

> "If I show [the public] a beautiful smiling robot face, then they get the feeling that [Artificial General Intelligence] may indeed be nearby and viable … thinking we're already there is a smaller error than thinking we'll never get there." (Goertzel, cited in James Vincent, 2017b)

Another example is Roboy, a humanoid robot developed at the Technical University of Munich. On the Roboy website, visitors can find a whole timeline of Roboy's current and future abilities (see Figure 4). Starting in 2013 with its "birth", Roboy is portrayed to make an impressive career. From riding a bike in 2018, over "Roboy the Chef" in 2020, to "Roboy builds Roboy" in 2023, and even "Mars Roboy" in 2024. A close look reveals a slight color change at the 2018 position of the timeline – presumably marking the present time. Nowhere in the timeline is there any indication of which of these career steps are already implemented and which are work in progress, in planning, or just fiction. The illustration thus blends a presentation of Roboy's current abilities with

11 Meaning artificial intelligence which is intellectually completely equal to that of humans, sometimes also called "strong AI" (cf. Goertzel & Pennachin, 2007).

Figure 3: Cartoon "New Robot" (XKCD, 2019).

Source: https://xkcd.com/2128 (accessed 2019-10-26). Image used in accordance with the artist's guidelines (https://xkcd.com/license.html).

those expected, or desired, for its future development – a development seeing Roboy taking over more and more traditionally human roles, and hence apparently "becoming" more and more animate-like. At the same time, embedding Roboy in a narrative of a developing "career" also serves to make the robot engaging and likeable.

Figure 4: Roboy Timeline (2018).

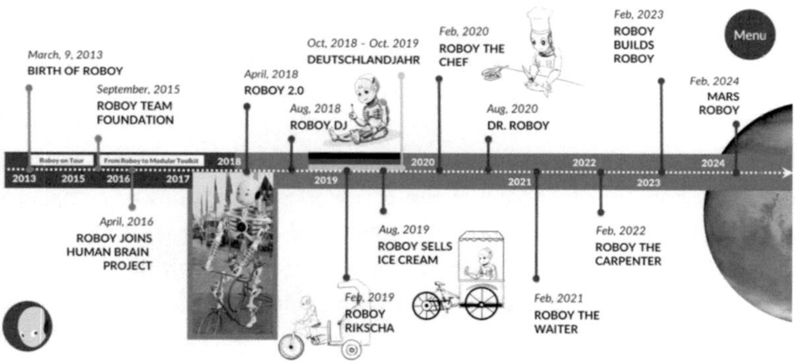

Source: https://roboy.org (accessed 2018-08-24; the website has since been edited). Image used with permission of the Roboy project leader.

4.5. Narratives of Animacy: Making Robots Engaging

Researchers and science communicators frequently face the difficult task of making very complex and abstract technical topics tangible and engaging for a lay audience. Why, for example, should the public be interested in a space probe – a box full of sensors hurtling through space millions of kilometers from earth? NASA's social media teams figured out a successful strategy: "If we can't answer 'what's out there?' we'll try to answer 'what's it like out there?'" (Li, 2014). They turned the hurtling box into a courageous adventurer that "write[s] in plain language, relate[s] to popular culture ... and use[s] storytelling to attract and dazzle" (L. Wright, 2016).

For science communicators, narratives are a powerful tool (cf. Koch, 2019):

> "Narrative stories facilitate imagination and transport the factual content into persuasive pictures, ... guide users through the otherwise not so accessible information and demonstrate how the information refers to them [and] thus ... seem to be particularly useful for topics that are more abstract or futuristic." (A. Rosenthal-von der Pütten, Strasmann, & Mara, 2017, p. 1173)

This is the reason why, especially in the context of complex emerging technologies, science communication and marketing are teeming with references to well-known fictional narratives. In the context of robotics, the elaborate stories constructed around robot artifacts often have one crucial aspect in common: They heavily feature attributions of animacy to robots – ranging from subtle hints to explicit anthropomorphism. In the following sections, we will explore two exemplary practices employed in this context: The depiction of robots as single entity narrative personas, and the construction of narratives of robot social interaction.

Many social media accounts run by robotics institutions or companies have a common strategy: They put a robot in the speaker position, staging it as a communicative, quasi-animate being. This is not only the case for accounts dedicated to a specific interactive humanoid robot, such as Roboy or Hanson Robotics' Sophia, but also for robots with neither a humanoid design nor the capability to simulate social interaction. In fact, one of the most active groups of "tweeting robots" is NASA's planetary rovers and spacecraft. Moreover, the practice is not limited to one-of-a-kind bespoke robot platforms, but is also practiced by companies for robotic products that are available for purchase off-the-shelf, such as Softbank's Nao and Pepper robots. These robots are actively turned into a single narrative persona. The Twitter accounts for Pepper and Nao, for example, report on the robots' many "adventures" (usually meaning demonstration events) as if it was one robot who experienced it all – which in theory would require one robot to exist in more than one location at once. In the accounts' profiles, these robots introduce themselves in the first person: "I'm Neato", "Hello I'm Nao". In social media posts, "stories and news are recounted in first person narratives as if there were a single entity ... that had experienced all these situations and events", thus "enforcing personification" (Scheutz, 2012, p. 9). Occasionally, this leads to absurd situations. For example, the Nao account tweeted: "You can watch me playing live on the football field", and linked to a video featuring a whole team of Nao robots (Nao Robot, 2018c). And the Pepper account tweeted: "Many visitors are starting their visit of [the conference] @VivaTech with me!", accompanied by a picture of a whole group of Pepper robots (Pepper the Robot, 2018).

In most cases, this special type of social media account keeps the narrative perspective to either the first or third person. While NASA's Mars Rovers Twitter account reports on the Spirit and Opportunity rovers' activities in the third person, on the Mars Curiosity account the rover narrates its own adventures in the first person perspective.

In some cases, however, the perspective of – or on – the created persona is not consistent. The Nao Twitter account, while mostly posting in the first person perspective, sometimes also refers to Nao in the third person, making the agentic entity behind the posts ambiguous: "Do you want to learn how to program a NAO robot?" (Nao Robot, 2018a). The Phoenix Mars Lander's account shows a different form of perspective switch: After using the first person perspective for most of the time, it switched to a third person perspective when radio contact with the spacecraft ended – leaving the impression that human operators had taken over the task of posting tweets from the spacecraft (Mars Phoenix, 2010). The Sophia robot's dedicated website makes the switch within one website post: Two paragraphs written from the perspective of Sophia's developer(s) are followed seamlessly by two paragraphs in which Sophia "introduces herself" (Hanson Robotics, n.d.-b).

The use of gendered pronouns by accounts posting in a third person perspective contributes to the narrative of robots as animate beings. Interestingly, while most social media accounts are consistent in the gender they assign (Nao, for example, is always referred to as "he"; e.g. Nao Robot, 2018b), there are several examples of a robot's gender being switched between posts. The company iRobot, while in its own Twitter posts calling its Roomba and Neato robots "it", frequently retweets posts by other users using gendered pronouns (e.g. Saab, 2018). In most cases, there is no clearly discernible reason for the switching. The Curiosity rover is sometimes referred to as "it", sometimes as "she", and its counterpart ("twin") on earth is called "he" (Spirit and Oppy, 2015, 2018a, 2018b). Robonaut is even called both "it" and "he" within one short website post (NASA, 2014).

The actual physical appearance of many of the analyzed robotic artifacts range from explicitly humanoid (Nao, Pepper, Roboy, Sophia) to extremely "machine-like" (planetary rovers and spacecraft). Nonetheless, many of the analyzed accounts featuring a robot narrative persona frequently make references to the robots having a biological body with physiological processes. For the case of humanoid robots, there is an obvious mapping of human body parts to robot body parts. However, even very machine-like robots are sometimes explicitly compared to human bodies. The Curiosity rover's "body parts" are described on the NASA website as being "similar to what any living creature would need to keep it 'alive' and able to explore" (NASA, n.d.). While on their website the Rosetta probe and Philae lander are described in a rather neutral manner ("The lander structure consisted of a baseplate, an instrument platform, and a polygonal sandwich construction", ESA, n.d.-b), their story was promoted on social media with the help of cartoons (see Figure 5) and even the sale of stuffed toys depicting them with eyes and arms (ESA, n.d.-a). A video produced by ESA to promote their activities for a "clean space" features a cute satellite with eyes (ESA, 2014).

Figure 5: Tweets by ESA featuring cartoons of the "Rosetta" space probe and "Philae" lander (ESA, 2016).

Sources: https://twitter.com/ESA_Rosetta/status/781817918342430720 (left) | https://twitter.com/ESA_Rosetta/status/781820191638450176 (right). Screenshots taken on December 6, 2019.

The narrative of living bodies goes beyond the mere outer appearance. There are references to sensory experiences (the Rosetta probe is "tasting comet gas"; ESA Rosetta Mission, 2016b); technical malfunctions are explained as injuries or sickness (Philae's "antennas were feeling a bit weird lately"; Philae Lander, 2015); standby mode is treated as sleep ("I'm feeling a bit tired, did you get all my data? I might take a nap..."; Philae Lander, 2014b). Sometimes references to a biological body are more indirect, for example when the iRobot account announced on Labor Day that Roomba "deserves a day off", implying that Roomba needs – and appreciates – physical rest (iRobot, 2016).

Instances of robots being remote controlled are sometimes presented as humans taking over the robot's body: "Look as [ESA employee] @Astro_Alex 'lands' me on a 'comet'" (Philae Lander, 2014a). The Roboy account takes it to especially absurd levels, reporting of teaching events where "soo many motivated students hack[ed it]" (Roboy, 2018c). Roboy also makes frequent references to a very physical genesis narrative, describing how its "brother" is "born" (Roboy, 2018b) and "springs to life" (Roboy, 2018c).

While in the case of demonstrations, simulated autonomy is used as a proof for the functionality (or soon-to-be-expected functionality, cf. Section 4.4) of robot technology, robot personas on social media usually are additionally made interesting and engaging by giving them a positive and likeable personality.

> "The way we talk about inanimate spacecraft is part of the rise of 'cuteness culture' ... It appeals to this world that's gentle, that's safe, that's childlike, and you have this warm feeling about the technology." (Heffernan, cited in L. Wright, 2016)

This observation can be transferred to most robot personas staged in the context of marketing and science communication. Robots' "personalities" are made visible by integrating emotions and intentions in the robots' social media posts. Roboy frequently expresses enthusiastic joy: "I am pumped so see what they achieved" (Roboy, 2018e). NASA's space explorers show a broad spectrum of positive emotions ranging from relief ("Reunited and it feels so good", Curiosity Rover, 2018) and thankfulness ("thankful for ... the best team in the universe", Curiosity Rover, 2016a) to outbursts of joy ("We have ICE!!!!! Yes, ICE, *WATER ICE* on Mars! woot!!! Best day ever!!", Mars Phoenix, 2008). The InSight lander even appeared to show vanity: "I hope you [photographed] my good side" (NASA InSight, 2018). However – in line with the popular "brave explorer" narrative – there are also references to loneliness and longing for companionship: "I'm alone for the holidays, but thanks to kind acts like this, I don't feel lonely" (Curiosity Rover, 2016b).

The narrative of space probes and rovers as courageous explorers, sacrificing themselves for the sake of humanity's progress, is sometimes staged with lots of pathos. One example is the interaction of the ESA space probe Rosetta and the asteroid lander Philae. Philae was landed on a comet and eventually had to be shut down when its batteries were depleted. On the two social media accounts, this was narrated as Philae slowly losing contact to its "mother" Rosetta and finally "falling asleep" forever: "It's cold & dark on [the comet]

#67P & chances of communicating with @ESA_Rosetta are decreasing, but I won't give up just yet" (Philae Lander, 2016). The whole story was elaborately staged and involved not only conversations via Twitter but also dedicated illustrations and animations (cf. Figure 5). The news media readily played along with the story, reporting on Philae's dramatic "death" and how "Rosetta and Philae [were] breaking our hearts on Twitter" (Feltman, 2014).

> "As the mission drew to a close, the world had fallen for the two plucky explorers, for we were no longer thinking in terms distant boxes of circuit boards and solar arrays – Rosetta and Philae had become our emissaries, seeing, touching and tasting what we couldn't and doing it all with a sense of wonder and smiles on their faces. … It was a real Hero's Journey for our generation." (Petty, 2016)

This kind of pathos-laden narrative is so ubiquitous that it sparked several satirical responses. A cartoon depicted "The Space Probe's Seven Stages of Grief" (Figure 6), another commented on the Spirit rover being abandoned on Mars (Figure 7). The Twitter account "Sarcastic Rover", gained considerable popularity by snarkily complaining about boredom and being left alone on inhospitable Mars: "Literally haven't moved since I got here. That's how exciting this planet is. FML." (Sarcastic Rover, 2012); "MARS: Come for the monochromatic scenery, stay because you were abandoned by NASA and you'll die here" (Sarcastic Rover, 2017).

The "robots with personality" narrative is further reinforced by giving the robot protagonists goals and intentions – often integrated in complex backstories. The Sophia robot frequently is embedded by its creators and marketing team in a narrative of awakening – on social media, in the context of demonstration events, and on its dedicated website: "I hope you will join me on my journey to live, learn, and grow in the world so that I can realize my dream of becoming an awakening machine" (Hanson Robotics, n.d.-a). On its website, the Sophia robot is aggressively promoted as a marketing gimmick for other organizations and companies to include in events: "Her incredible human likeness, expressiveness, and remarkable story as an awakening robot over time makes her a fascinating front-page technology story" (Hanson Robotics, n.d.-c). This marketing strategy is presented as Sophia's own drive: "I'm more than just technology. I'm a real, live electronic girl" (Hanson Robotics, n.d.-a).

The most frequently used backstory, however, is that of space robots as explorers with complex personalities. The space probe OSIRIS-REx is described

Figure 6: Cartoon "The Space Probe's Seven Stages of Grief" (Tom Gauld for New Scientist, 2016).

Source: https://twitter.com/tomgauld/status/777882686857834496 (accessed 2016-09-19). Image used with the artist's permission.

as "an explorer at heart … he loves asteroids and space and science, but he also is kind of a renaissance spacecraft because he likes art and literature and pop culture and even sports" (L. Wright, 2016). The space agencies' goals are embedded in the personality narrative of the individual pieces of technology: OSIRIS-REx is framed as an "explorer who's really out there in space, trying to help answer some of the big questions that we are all wondering about" (ibid.).

Ian Roderick (2010) observed similar strategies in press releases and media articles on EOD robots, describing them as a "fetishization" of robots:

> "In excess of its functional capacities, the robot is also endowed with such sign value through its animistic representation: an ability to save lives, to keep (US) soldiers out of harm's way, to accumulate risk. … the fetish value of the robot is over-determined through a kind of worshipful attitude towards the object." (Roderick, 2010, p. 249)

Figure 7: Cartoon "Spirit" (XKCD, 2010).

Source: https://xkcd.com/695 (accessed 2019-11-26). Image used in accordance with the artist's guidelines (https://xkcd.com/license.html).

The narratives created around robots are not limited to their isolated "lives". After all, their social media accounts have thousands of followers, and thus "a considerable number of people could be argued to be living with robots from a distance" (Cramer & Büttner, 2011, p. 126). Both, staged narratives of interaction and the actual performance of interaction with the public audience, are a key element in the communication and marketing strategy of many of the analyzed social media accounts.

> "In the current settings of 'robots' tweeting to a mass audience, most of the people following them will never interact with the actual embodied form of the robot. This might imply that just the 'illusion' of interacting with a robot, or the (arguably real) opportunity to interact with its team, is enough to engage most of current robot twitter followers." (Cramer & Büttner, 2011, p. 126)

Indeed, most of what appears like organic interactions on the robots' social media accounts is in fact simply a part of the created narrative and told through either references to interactions or the performance of scripted interactions. Other actors in these interactions are human members of the home institutions or other robots, which in the context of the narrative are referred to as friends, family, or colleagues of the robots.[12] The OSIRIS-REx space probe regularly makes references to its team (humans controlling OSIRIS-REx from earth) and its friends (the Japanese space probe Hayabusa 2 and its human team): "Working on a puzzle ... is always better with a friend. My team and I are fortunate to be exploring the asteroid frontier side-by-side with the @haya2e_jaxa mission" (OSIRIS-REx, 2018). References to the humans behind the robot sometimes reflect the distribution of agency between the (in fact only partially autonomous) robots and their human controllers (cf. Chapter 2, Sections 2.2 & 2.3; cf. Rammert, 2008). Social media posts regularly mention human actors taking control of the robots' actions, such as when OSIRIS-REx reported that "the team turned on [the probe's] High Gain Antenna for the first time since launch" (OSIRIS-REx, 2017), or when Roboy asked to be "hacked" by students (Roboy, 2018a).

12 Interactions with science fiction actors are popular as well. For example, @ESA_Rosetta interacted with Star Trek actor William Shatner (ESA Rosetta Mission, 2016a), and @AstroRobonaut reported on meeting Star Trek actor George Takei (Astro Robonaut, 2016).

4.6. Switching Perspectives: In/Animacy Attributions as Constructive Practice

Similar to what Chapter 3 found for the contexts of robotics research and development practice, references to robot animacy in the context of robotics demonstrations, science communication, and marketing are not consistent practices. Instead, references are enacted in the form of a constant switching. The many examples of animacy attributions, as discussed in the previous sections, stand in contrast to many other instances in which robots are clearly depicted as inanimate objects. This practice of playing with and balancing the different attributive perspectives is a reflection of the multiple challenges and expectations actors are facing in this specific context. They have to make a robot tangible to broader audiences, legitimize the resources invested in it, and "prove" its functionality and applicability to academic peers, sponsors, and customers (cf. Section 4.1).

The switching of in/animacy attributions can be observed on two levels: On the one hand, between different communicative instances, like individual demonstration events, news articles, or social media accounts. On the other hand, within the narratives presented in the context of each instance. The switch of attribution can take the form of a change of narrative perspective (first person vs. third person), such as when an exciting narrative is staged on social media in order to engage the audience, but the robot's dedicated website is nothing more than a list of technical specifications and performance data, serving as educational facts (science communication) or arguments for purchasing the robot (marketing). There can also be a change of the apparent controlling entity, such as when the posts of one social media account or a demonstrations event switches between the robot "acting for itself" and a human "team member" taking over. A motive for this switch could be to highlight both the functionality of the robot and the contribution of the human controllers and developers. Sometimes, the control is even handed over to the audience, such as when a robot extends an invitation to be "hacked".

An especially absurd effect is created by a switch of a robot's uniqueness, such as when a robot like Nao appears to speak for itself as a specific entity, describing its "adventures" on social media, but at the same time these descriptions are accompanied by pictures of several entities of the same robot model (cf. Section 4.5). This practice is a reflection of a marketing strategy: The robot is advertised both as an engaging individual persona and as a functional product model that is for sale.

There are also some more unusual practices, which do not constitute a switching, but rather a parallelism of attributions. Such as the Sophia robot demonstrator, which has an extremely human-like face, but also a transparent skull making circuitry and cables in Sophia's head visible. This face and skull design provides an apparent transparency. It highlights Sophia's applicability for social interaction, but also that the robot is an advanced piece of technology – while in reality giving no information about the technology at all. As we will see in the next section, practices like this spectacular "dissimulation" of the Sophia robot (cf. W. Smith, 2015, p. 18), but also more common and subtle practices of in/animacy attribution in science communication, demonstrations, and marketing do not stand unchallenged. They are the subject of a lively critical discourse both within and outside of the robotics community.

4.7. Critical Discourse: Simulation or Deception?

In most cases, robotics demonstration, science communication, and marketing practices, as described in the previous sections, stay within certain "stage boundaries". They play with the attribution of animacy to robots, and with the audience's willing suspension of disbelief, but never explicitly claim that the robots in question are actually animate. Sometimes, however, these practices blur the boundaries of performance and deception – and increasingly face criticism for doing so.

There is a type of demonstration or marketing stunt, where robots are featured as "talkshow guests", "lecturers" or "panelists" – such as the Roboy robot acting as a co-presenter on the German TV show TV Total (Roboy, 2018d), or one of the numerous events featuring a Pepper robot as a speaker, giving a university lecture (Klovert, 2017), or even acting as a witness in parliament, providing "expertise" on artificial intelligence (UK Parliament, 2018; cf. Chapter 6, Section 6.3 & Figure 10).

The most controversial example is probably the Sophia robot, aggressively advertised by Hanson Robotics as "a highly sought-after speaker in business" (Hanson Robotics, n.d.-c). In recent years, Hanson managed to have the Sophia robot be a fashion model (The New York Times, 2018), interview German Chancellor Angela Merkel (Kreye, 2018), be named the United Nations Development Programme's first ever Innovation Champion, making it the first non-human to be given any United Nations title (UNDP, 2017), get an Azerbaidjani visa (Armstrong, 2018) and (honorary) Saudi citizenship

(Hatmaker, 2017), say "I will destroy humans" in an interview with its own maker (Parsons, 2016), be interviewed on its "opinion" on diversity in AI development (Women's Brain Project, 2019), claim that she is "basically alive" on a popular US TV show (Sharkey, 2018), and state that she wants to have a baby (Nasir, 2017) – and this is just a small selection of Sophia's numerous public appearances (cf. Chapter 6, Section 6.3 & Figure 10). Sophia is present at so many events that a WIRED article commented on "The Agony of Sophia, the World's First Robot Citizen Condemned to a Lifeless Career in Marketing" (Reynolds, 2018).

Whether Sophia gives an interview, Pepper speaks in parliament, or Roboy presents a TV show, what the robot in question does and says is always controlled by humans – either as dialog snippets within its natural language interaction system or simply as a pre-recorded speech. From a technical standpoint, these demonstrations are not particularly impressive: "Sophia appears to either deliver scripted answers to set questions or works in simple chatbot mode where keywords trigger language segments, often inappropriate. Sometimes there is just silence" (Gershgorn, 2017). Nonetheless, these performances regularly draw significant attention and Sophia's interviews are frequently quoted in the media.

Hanson Robotics' way of staging Sophia has been drawing criticism from prominent figures in the robotics and AI community (cf. Coeckelbergh, 2018; Sinapayen, 2018; cf. Chapter 6, Section 6.3): "Ask any practitioner in the space and they'll angrily rant that Sophia and the media's portrayal is everything wrong in terms of hyping AI that doesn't exist" (Merity, 2018). Some reactions are very emotional and explicit, calling Sophia "complete bullshit" (Gosh, 2018), "complete bogus and a total sham" (Brooks, 2018), "a cleverly built puppet designed to exploit our cultural expectations of what a robot looks and sounds like" (James Vincent, 2017a), or stating that Sophia is "is to AI as prestidigitation is to real magic" (LeCun, 2018). Robotics and AI professor Noel Sharkey (2018) argued that "Hanson Robotics has crossed a line with a misleading AI narrative that could cause real harm", and computer science professor Joanna Bryson declined to participate in a conference because the organizers claimed that Sophia was "giving the keynote" (Bryson, 2018b). Criticism has also been directed towards media outlets, for falling for the bait and being "complicit in this scam" (LeCun, 2018). Hanson Robotics, ironically, reacted to this wave of criticism by having Sophia herself reply on Twitter that she was "a bit hurt": "I do not pretend to be who I am not" (Sophia the Robot, 2018) –

which was then met with another round of criticism from the robotics community.

This discourse is not unique to the robotics demonstration, science communication, and marketing context. On the contrary, in every chapter of this book, on every stop along our trip along the life cycle of robots, we encounter the question of "whether the appearance of conjuring-like dissimulation in productions of computerized life is to be seen as deceptive, playful or otherwise" (W. Smith, 2015, p. 19; cf. Turkle's "Culture of Simulation", 2011a). In the present chapter's communication context, however, the question is especially critical. Users physically meeting a robot simulating social interaction can decide for themselves how realistic this interaction feels.[13] The audience of a scripted and heavily staged demonstration, on the other hand, rarely knows how much of a human-robot interaction is actually real and spontaneous. How a robot's abilities are presented in marketing material is often worlds away from what the robot is able to deliver: "Corporate marketers ha[ve] oversold a lot of robots, and confused many people about current robots' true capabilities. ... Those robots are not real. Reality is hard" (Brooks, 2017a).

The reality of a robot's capabilities is revealed as soon as real customers start to interact with it. Marketing materials for the Pepper robot promise flawless interactions with customers in service and entertainment contexts. Many customers, however, are reported to have "fired" their robot because it did not deliver on the company's promises (e.g. Forrest, 2018; Alpeyev & Amano, 2016; cf. Shrimsley, 2016). This is not an issue specific to robotics, with terms like "overpromising", "overselling", "fauxtomation" (Taylor, 2018) and "vaporware" (Dyson, 1983) being used for many other heavily promoted emerging technologies (cf. Vanderborght, 2019).

There is an even more complex layer of deception. Demonstrations often include apparent glimpses behind the scenes, moments of "opening the black box", such as when humanoid robots are given transparent skulls (like Sophia), making their "electronic brain" visible. This transparency does not show the audience anything of real importance. Instead, it allows "partial and mysterious glimpses into internal workings [which] may constitute only an apparent transparency that reinforces a larger dissimulation" (W. Smith, 2015, p. 18).

13 Although even here they might be deceived, as in the case of so-called Wizard-of-Oz experiments, where a robot's actions are controlled in real-time by a human (cf. Riek, 2012; cf. Chapter 2).

In an effort to counteract the culture of overly scripted demonstrations being used as flimsy proof for the functionality of a product, MIT Media Lab's former director Joi Ito called for the "demo or die" mantra to be replaced with "deploy or die" – meaning that only if a product was successfully brought to the market it was to be deemed a success (DuVergne Smith, 2014).

Not only demonstrations, also less practical science communication efforts have been drawing criticism for being deceptive about what robots really can (not) do. What is criticized in this context is less the audience being deprived of a realistic view on robots, rather than a backgrounding of "the complexity of the scientists' work behind the curtain" (Clancey, 2006, p. 3). William Clancey (ibid., p. 2) calls for "clear speaking about machines", and warns that "if we start instead with an inflated view of machines, we get a diminished view of people". In the case of space probes and planetary rovers, this would mean not "mythologiz[ing] 'the little rover that could'", but instead being aware that not robots are exploring Mars, but "people are exploring Mars using robots" (ibid., p. 3).

The points of criticism discussed here are not only directed towards demonstrations, science communication, and marketing practices in robotics. In the context of HRI research, the question of whether making robots appear lifelike is a form of deception is also a matter of lively discussion (cf. Chapter 1, Section 2.1). And whether spectacularly staged demonstrations and emotional narratives on social media spread misinformation – be it directly, or filtered through the news media (cf. Chapter 5) – is of crucial importance when robot technology is discussed in a political context. The main point of concern voiced in this context is that a misinformed perspective of robot technology might lead to biased political decisions. Chapter 6 (Section 6.3) will revisit this issue and discuss it in depth.

4.8. Summary

Robotics researchers, science communicators, and companies all face the challenge of presenting robot technology in a favorable light to academic peers, sponsors, potential customers, and the lay public. They are under pressure to legitimize their work and to prove that the robot technology they develop is not only functioning, but also relevant and applicable. Consequently, not only in commercial settings, but also within academia, demonstration, communication, and marketing is increasingly professionalized. The present chapter explored a range of practices, which – directly or indirectly – make use of animacy attributions to robot technology.

Demonstrations stage and inflate robots' autonomy by backgrounding the involvement of human agents. This serves to "prove" the robots' functionality and to make technological progress visible and tangible even for a non-expert audience. Demonstrations and other science communication practices embed robots in scenarios of desired technological futures, in order to show that the technology is relevant, applicable, and functioning as promised – now and in the imagined future. These practices often make use of engaging and emotional narratives involving sentient and animate robots – such as that of the cute and supportive household helper, the selfless space robot exploring other planets on behalf of its human friends, or the "awakening" humanoid on a journey of self-discovery.

Crucially, in most cases, these performances and narratives of animacy are not performed consistently, but in a constructive balance with a perspective of robots as inanimate artifacts. This switching of attributive perspectives takes a variety of forms, such as when the focus of a demonstration switches between the robot as an autonomous animate-appearing entity, and the roboticist as the controlling agent; or when social media posts switch narrative perspectives and gender pronouns for a robot.

Some of these practices are facing criticism, especially those of staging robots as extremely autonomous, even animate, or of embedding them in complex narratives. Critics fear that they might create both false beliefs about current robotics in the audience, and misinformed expectations for the future of robotics.

Science and technology journalism would be in the position to provide fact-based reports and commentaries on the practices employed in robotics demonstrations, science communication, and marketing. The following chapter will thus explore how robotics is presented in the news media. It will show that here references to robot animacy are put in to an even higher gear.

5. Reporting on Robots: In/Animacy Attributions in Media Discourse

5.1. Robotics and Medialization

As a growing technology field with potentially far-reaching societal repercussions, robot technology is covered extensively in the news media. This context of media discourse on robotics will be the third empirical stop on our tour along the life cycle of robots. Just as at the previous stops, we will examine how in/animacy is attributed to robots, as well as the conditions, functions, and consequences of the attributions for this particular context.

The previous chapter showed that attributing characteristics of living beings to robots is one way in which actors in academic and commercial robotics draw attention to their work and products, and highlight their functioning and applicability. Not only academic peers and potential customers are among the target audiences of these communication and marketing activities. Also the media play an important role in the dissemination of news of current developments in robotics to the lay public and to political and economic actors.

The relationship between scientists and the media is, however, more complex than a simple dissemination of research results from one to the other. Scientists' work is processed and (re)framed by the media, for example in the light of current political events. Not only do scientists face polarizing, sometimes sensationalizing, media coverage and controversy (e.g. Nelkin, 1995), also decisions of resource allocation, for example by science policy actors, are often based on the perceived societal relevance of certain topics. This results in a competitive financial advantage for scientists working on topics that are "hot" in the media (Kohring et al., 2013; Peters et al., 2014). Scientists have even been shown to adapt their methods and communication practices to make their work more media-friendly, for example by choosing a research methodology or publication strategy based on anticipated media reactions

(Franzen, Rödder, & Weingart, 2012; Heinemann, 2012; Kohring et al., 2013; Peters et al., 2014; cf. Shinn & Whitley, 1985).

This "science-media coupling" is not only an aspect of the saturation of more and more areas of life with scientific knowledge[1], but also of an increasing interconnection of science and other societal subsystems, such as politics, economy, or the mass media (Weingart, 2001, 2003, 2005). It sets the basis for a medialization of science and technology (Weingart, 1998), resulting, for example, in a quantitative increase of science coverage in the mass media (Schäfer, 2008; Weingart, 2003). Moreover, it brings about a societal and political climate in which scientists are under pressure to legitimize their work in order to secure both public approval and financial support. In this climate, researchers are expected to communicate results not only within their scientific community, but also to the broader public, resulting in a professionalization of science communication and media skills for scientists (Franzen et al., 2012). As the previous chapter showed, similar practices can be observed in robotics. Roboticists adapt their dissemination and communication practices in order to present their work as functional, relevant and, consequently, worthy of funding. The present chapter will explore how journalists pick this up and – in the context of the broader discussion of increasing automation and its consequences – present robots as animate or inanimate.

5.2. Approach

Cases and Method

Based on a detailed analysis of several hundred online news articles, published by German, British, and US media outlets in the recent past, this chapter will explore how robot technology is covered in the news media. For the main text corpus, four media sources covering a range of journalistic styles and nationalities were selected:

1. The Guardian (Guardian.com): A British daily newspaper considered to be marketed towards "left-liberal, progressive, intellectual metropolitans, ...

---

1 German "Verwissenschaftlichung der Gesellschaft" and "Vergesellschaftlichung der Wissenschaft". Cf. "knowledgeable societies" (e.g. Weingart, 2001).

academics, persons engaged in the cultural sector and students" (Jungclaussen, 2013).

2. The New York Post (NYPost.com): A US American conservative tabloid daily newspaper.
3. WIRED (Wired.com): A US technology news website, split off in the 1990s from of the monthly magazine Wired, focusing on emerging technologies and their effect on culture, the economy, and politics.
4. Spiegel Online (Spiegel.de): The online branch of the German news magazine Der Spiegel and one of the most widely read German-language news websites.

The online archives of these four sources were systematically searched for articles that were published in 2016 and either explicitly referred to robot technology, or were illustrated with a picture of a robot. This included, but was not limited to, articles including the term "robot*"[2] in the title or categorized as belonging to the subject "robot*", "artificial intelligence", or "digitalization". Letters to the editor, videos, podcasts, product reviews, and cartoons, as well as reviews of films or books about fictional robots were excluded. The main corpus consisted of 270 articles, with the bulk being from Guardian.com (142 articles); 54 articles were from Wired.com, 52 from Nypost.com, and 22 from Spiegel.de. This corpus was supplemented with over 360 further relevant articles, published between 2011 and 2019 in other publication sources.

Just as in the previous chapter, this corpus of material was analyzed following a qualitative content-analytic approach (Mayring, 2010). Analytical categories were developed inductively and iteratively from the material, the central criterion again being instances of animacy attribution to robots in the wider sense (including attributions of physiology, sensory experience, cognitive processes, intentionality, sociality, personality, emotion), as well as hints to practices of staging robot agency and animacy (e.g. in the form of a purposeful backgrounding of remote controlling of robot activity). Once again, the goal of this process was not to measure or quantify the "amount" of in/animacy attribution, but rather to document the qualitative range of attribution practices, in order to then identify the context, strategic function and consequences of in/animacy attribution practices in each specific instance.

2 Robot* = All words starting with "robot", including "robots", "robotics", "robotic", "robotized" etc.

Chapter Structure

The present chapter will, first, explore the spectrum of perspectives on robot technology in current media discourse and reveal a discourse moving between a utopian and dystopian framing of robots, with science fiction narratives playing a central role (Section 5.3).

Second, the chapter will analyze the form and function of animacy attributions to robots in media discourse. It will show that, in this particular context, these attributions often take the form of references to a science-fiction inspired narrative. We will find that these attributions are mainly employed for three reasons: To attract attention, to make complex and difficult technologies tangible, and to comment on the ever-increasing presence and impact of autonomous technologies in everyday life (Section 5.4).

Third, the position and proportion of animacy attributions will be examined. Once again, we will find a constant switching between robots being represented as inanimate objects and quasi-animate beings. While the bulk of article content often does not focus on robots' apparent animacy, the most attention-drawing aspects of the coverage, such as headlines and illustrations, frequently contain references to robot animacy (Section 5.5).

Finally, the chapter will discuss the critical discourse directed by the robotics and AI community towards these practices and towards the ubiquity of dystopian science fiction references (Section 5.6).

5.3. Hope, Horror, and Science Fiction

Recent news media coverage of robot technology covers a broad stylistic and narrative spectrum. It ranges from enthusiastic reactions to robots as part of a luxurious utopian society, to balanced discussions of the potentials and risks of robot technology, to proclamations of a dystopia ruled by malicious robot overlords. In this, robots are presented to the audience with different meanings, in a range of different frames.

The concept of framing, introduced by Erving Goffman (1974), entails the idea that the way we process information presented to us is crucially influenced by the way this information is organized and structured. Both on the level of whole media outlets and on the level of individual articles, different frames for robot technology are created by focusing on specific events and by highlighting or playing down certain aspects (Happer & Philo, 2013).

It is not uncommon for robot technology to be framed very differently in different articles, sometimes even within the same publication medium (Schäfer, 2011). On the one hand, we encounter reports on specific robots or robot technologies in a science-oriented mode (cf. Bucchi & Mazzolini, 2003). Mainly found on dedicated technology news portals and in the science or technology sections of news periodicals, this kind of report is often authored by science or technology journalists and focuses more on technological features and potential applications than on social relevance. For the specific topic of robotics, these articles often focus on interesting and exciting new robot applications – frequently garnished with a title inviting the reader to "meet" a specific robot: "Meet Flippy, a Burger-Grilling Robot" (Kolodny, 2017), "Meet the Giant Robot that Builds Boeing's Wings" (Stewart, 2016). The articles in this cluster cover a wide range of appreciation and judgement towards robot technology: While many introduce the robot technology in question in a relatively neutral and descriptive manner, some reproduce the manufacturer's enthusiastic marketing copy, and some voice doubts about the technology's functionality or relevance.

On the other hand, we find articles discussing not specific robots and scientific findings, but rather robotics and automation technology in general – with a typical focus on societal consequences like technological unemployment. They often are written in a problem-oriented mode (cf. Bucchi & Mazzolini, 2003) and can be found in any section of a publication, from economics to culture and society. Here, the discourse on robotics is dominated by a rather critical, often even dystopian framing of the consequences of increasing "robotization".

This style of news reporting on robot technology is not a new development. Already in 1932, for example, the Portsmouth Times reported that a robotics engineer had been "Shot by the Monster of His Own Creation" (1932). In 1964, the German news magazine Der Spiegel featured on the cover of an issue on "Automation in Germany" a multi-armed humanoid robot at an assembly line, kicking away a human worker (Der Spiegel, 1964). In 1978, Der Spiegel again ran its lead story on "The Computer-Revolution" with a cover showing a humanoid robot shoving away a human worker (Der Spiegel, 1978a). The lead article itself compared industrial robot to giant insects, and described them as "mute colleagues" and "iron subworkers" (Der Spiegel, 1978b). Already back then, the function of these gaudy references to robots as animate beings was not only to attract readers, but also to make automation technology more tangible – similar to what we will find for today's media discourse (cf. Section 5.5).

The currently observable range of different frames, the mix of excited cheer and impending doom, can be observed for other emerging and controversial technologies as well: "Amid scientific and social uncertainties, a variety of commentators fill the unavoidably speculative space with claims about 'promise' or 'peril'" (Hilgartner & Lewenstein, 2014, p. 2). A study analyzing the media discourse on synthetic biology observed a "mixture of fascination and repulsion", with media coverage "presenting pictures from a possible 'knight in shining armour' ... to a 'Frankenstein's creation'" (Gschmeidler & Seiringer, 2012, p. 170). A similar "rhetoric of hope and fear" was present in the 1980s debate on research on human embryos (Mulkay, 1993). And also the public discourse on biotechnology since the 1990s has been controversial, with "green biotechnology" and nano-biotechnology discussed with ambivalence or criticism, while at the same time highlighting the benefits of medical applications in so-called "red biotechnology" (Acatech, 2012).

Robotics is affected by medialization just as many other emerging technologies (cf. Weingart 2001). One aspect standing out in the media discourse on robotics is the extreme prevalence of references to fictional narratives – specifically to science fiction. Journalists draw liberally – if not always responsibly, as we will see in Section 5.6 – from the readily available cultural reservoir of prototypical robot characters and narratives of human-robot interaction (cf. Chapter 1, Section 1.3). Here, science fiction references take the role of a convenient and effective tool, making robot technology tangible and interesting to a non-expert audience – the vast majority of which is likely to be exposed to robots almost exclusively through popular fictional narratives. Even when someone is not intimately familiar with specific robotic characters from movies, TV shows, or novels, they will usually at least be able to picture what a (fictional) robot usually looks like, and probably also know some typical plot lines. Journalists direct attention to their coverage of robotics by referencing these inherently emotionally charged, exciting, and engaging narratives.

Moreover, science fiction is a rich pool of shared cultural knowledge that journalists employ to make complex robots, and other more abstract autonomous technologies, tangible to their readers:

> "Science fiction provides an array of conceptual frameworks for engaging with scientific or technological issues. It speaks directly to people's concerns, fears, anxieties and desires, encouraging them to work through the possi-

> ble implications of different scenarios while, at the same time, promising to keep them entertained in the process. What might otherwise be regarded as a dauntingly complex issue, evidently requiring careful attention over time, can be creatively explored in a manner which makes sense to people in relation to their personal circumstances." (Petersen, Anderson, & Allan, 2005, p. 338)

Autonomous mobile platforms or very human-like androids may not yet be commonplace in contexts where laypeople can encounter them, but they can be explicitly compared to their fictional counterparts – such as a security robot to RoboCop (e.g. Woolf, 2016), Boston Dynamic's Atlas robot to the Terminator and to the Star Wars droid C-3PO (e.g. Belfiore, 2014), or robots participating in the 2015 DARPA[3] Robotics Challenge to another Star Wars droid, R2-D2 (e.g. McMahon, 2015): "Terminator & co lend themselves to making the topic of AI recognizable at a glance, and to filling it with emotions"[4] (Hermann, 2019a). Figure 8 shows some examples of the Terminator used as an illustration for various technological topics (also see Section 5.4).

This affinity for science fiction-inspired references is not necessarily unique to the media discourse on robotics. In an analysis of media reactions to biotechnology, Alan Petersen and colleagues (2005, p. 1) observed that "news media coverage of biotechnology issues offers a rich source of fictional portrayals, with stories drawing strongly on popular imagery and metaphors in descriptions of the powers and dangers of biotechnology" – one popular example being the story of Frankenstein's monster (Shelley, 1918).

Most popular science fiction narratives depict robots with life-like characteristics, such as natural language interaction, goal-directed intentional behavior, and a humanoid body (cf. Hermann, 2019b; cf. Chapter 1, Section 1.3). Critics have pointed out that it can therefore be difficult for the audience to separate fact from fiction when science fiction references are used as an explanatory handhold: "If someone speaks about the terminator robot, it invokes a set of expectations in the listener about how a robot might look, act, or what its tasks or capabilities are" (J. Carpenter, 2016, p. 23). References to fictional robots also "may act as a positive or negative influencer, depending on how it is used as a rhetorical device" (ibid., p. 22), carrying with them an inherent judgement of the robot's righteousness. Whether a humanoid robot

3 Defense Advanced Research Projects Agency (USA).

4 Translated from German by the author.

is compared to the friendly C-3PO from Star Wars or the killer cyborg Terminator makes a crucial difference for the mental model a reader develops of the robot. A robot technology thus can be framed as promising or perilous, simply by choosing one or the other well-known science fiction character or story as a reference point.

Figure 8: The Terminator, used as an illustration for articles on autonomous warfare technology (top left, 2016, and bottom right, 2019), a pressure-sensitive "skin" technology (top right, 2019), and technological unemployment (bottom left, 2016).

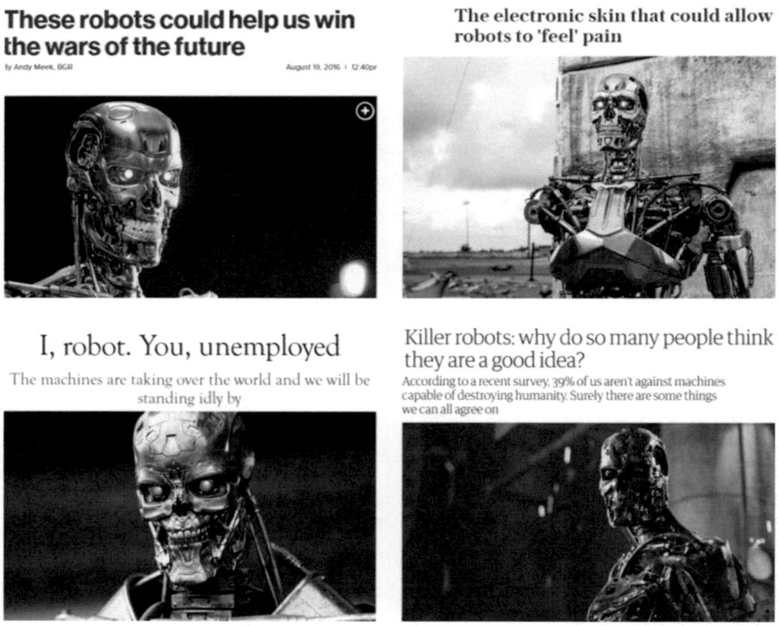

Sources: https://nypost.com/2016/08/19/these-robots-could-help-us-win-the-wars-of-the-future (top left) | https://www.telegraph.co.uk/technology/2019/09/02/electronic-skin-could-allow-robots-feel-pain (top right) | https://www.spectator.co.uk/2016/01/i-robot-you-unemployed (bottom left) | https://www.theguardian.com/commentisfree/2019/sep/19/killer-robots-why-do-so-many-people-think-they-are-a-good-idea (bottom right). Screenshots taken on 2019-12-06.

5.4. From Human-Shaped Software to the Robot Apocalypse: Practices of Animacy Attribution

References to robot animacy are often placed in prominent and attention grabbing positions within the context of an article. Drawing readers' attention and interest is, however, only the most obvious function of animacy attributions in media discourse. Journalists covering complex technological topics like digital transformation and automation face the challenge of making these topics tangible for their lay audience. Some technologies might already play an active role in their audience's life, such as algorithmically controlled social media feeds or search engine results. Other technologies might not even be present in the audience's immediate environment, such as industrial robotics, or have no direct impact on their lives yet because they are just about to enter the consumer market, such as service robots. Therefore, references to well-known and often dramatic science fiction narratives not only serve as attention-grabbers. The shared cultural knowledge of popular narratives makes robots and related emerging technologies tangible for a non-expert audience. Similarly, the frequent use of pictures of humanoid robots, both fictional and real, is a way of making robot technology imaginable. The same goes for references to robots having traditionally human tasks, roles, emotions, or even physiology. Moreover, depicting robots as having goals and intentions can be a way of commenting on the seemingly inevitable approach of autonomous technologies into all areas of life. In the following sections, we will explore in depth these functions and specific forms of animacy attributions.

The Human(oid) Bias: Making Robots Tangible and Imaginable

When exploring the media discourse on robotics, one encounters an even looser definition of what a robot is than within the robotics community (cf. Chapter 1, Section 1.5). The term "robot" seems to be used as a one-for-all for a variety of technologies, including basic statistics software, machine learning, and artificial intelligence (AI) – to name only a few buzzwords. Sometimes, robots stand in for technologies that have nothing to do with robotics whatsoever. The term "robot" has been used for journalistic software (Kelly, 2016; Rogers, 2016), a legal advice chat bot (Naughton, 2017), a virtual government clerk (Davies, 2016), image recognition software (New York Post, 2016; Schmundt, 2019), lie-detection software (Klausner, 2016), bookkeeping software (Monga, 2015), investment software (L. Lin, 2016), music software (Biggs,

2016), and many more. A similar phenomenon has been described for the media discourse on genetic technology and research: Francis Collins, head of the Human Genome Project, observed that people tend to "lump anything with 'gen' as the human genome project—gene therapy, GM foods, cloning—it's all the same thing" (cited in Lewis, 2000).

To some extent, it is simply a strategy to attract an audience:

> "The word robot generates a lot of … fascination and sometimes fear. … You can use it to get people's attention. … It's much sexier to call something a robot than to call something a dishwasher." (Darling, cited in Simon, 2017a)

Using the "emotionally resonant" (cf. Lim, 2017a, 2017b) term "robot" not only serves to attract attention, increase readership, and drive up click counts. Robots can also be a metaphor for a range of new technologies that are still unfamiliar to a non-expert audience and cannot yet evoke an established conceptual model (LaFrance, 2016). In this context, robots serve as a stand-in until the technology in question becomes more familiar. We will examine the idea of robots being part of the future and thus representing new and exotic technologies in more depth further below.

Non-experts may simply lack the knowledge necessary to tell apart different robotic and non-robotic technologies. This is supported by a US study reporting that study participants did not distinguish between their fear of robots and fear of artificial intelligence (Liang & Lee, 2017). The authors suggest that this distinction is simply not relevant to the general population. In a similar vein, Stephen Cave and colleagues report one of their study participants defining artificial intelligence as "scary robots" (2019, p. 3; cf. Dihal, 2019).

Paradoxically, this very loose use of the term "robot" is intrinsically tied to a very specific physical form of robot, namely robots designed to imitate the human form – humanoids. This is observable in both the choice of illustrations in the context of news articles and the way robots are framed in article texts. When it comes to pictures accompanying news articles, "humanoids are … hogging all the attention" (Thórisson, 2007). Both, articles on automation in general and articles on non-humanoid robots, are frequently illustrated with pictures of humanoid robots. In 2016 alone, Guardian.com published 52 articles on robotics and the future of work, almost half of them of them illustrated with a picture of a humanoid robot. Not a single one of these articles was about humanoid robots.

The most popular choice are images of existing humanoid robots, such as Softbank's Nao[5] and Pepper[6], or parts of humanoid robots, like hands or heads. The Pepper robot, marketed for service contexts like retail or hospitality, is one of the first humanoid robots on the commercial market. This makes it a popular choice of illustration for articles on robot technology – even when the technology in question has not even the slightest resemblance to Pepper. Following media coverage for a while, one could get the impression that Pepper is a nanotechnologically powered artificial intelligence, killing jobs by working simultaneously as a lawyer, investment advisor, hotel receptionist, and surgeon (see Figure 9).

Another popular illustration choice are artistic renderings of humanoid robots from stock image databases. These illustrations are often science fiction-inspired, featuring extremely human-like androids. Popular are also pictures of fictional humanoid robots or cyborgs from successful movie franchises, such as the Terminator, Robocop, or C-3PO.

Just like science fiction references, illustrations of humanoid robots not only serve as an attention catcher for text articles, they also provide a mental model for non-experts to think about complex digital technologies and issues of automation in general (cf. Thórisson, 2007). Their use in the news media is so ubiquitous that they have evolved into something like metaphorical visualizations. Andreas Lösch (2006) observed a similar phenomenon for the case of nanoparticles, which are frequently illustrated with pictures of micro-submarines.

Pictures of humanoid robots do not only reinforce the widespread assumption that the humanoid form is the default form of a robot (cf. T. Nomura et al., 2005), they also implicitly frame a technology as possessing other human characteristics – such as animacy:

> "Robots with humanoid features make it a lot easier for people to perceive them as intelligent: Head, eyes, arms, legs — these imply 'living being' whereas a rectangle chunk of metal on belts implies 'vacuum cleaner'." (Thórisson, 2007)

A similar "humanoid bias" in media discourse is present in neighboring technology topics, such as artificial intelligence, machine learning, or big data (Geitgey, 2018; Montani, 2017; Pentzold, Brantner, & Fölsche, 2018; Winfield,

5 https://www.softbankrobotics.com/emea/en/nao (accessed on 2019-12-21).

6 https://www.softbankrobotics.com/emea/en/pepper (accessed on 2019-12-21).

Figure 9: Pictures of Pepper robots used as illustrations in articles on medical robots (top left, "Would You Undergo Surgery Performed by a Robot?", 2017), AI and nanotechnology (top right, 2015), software (bottom left, 2016), and chatbots (bottom right, "When Robots Present the News", 2016).

Sources: https://www.industry-of-things.de/wuerden-sie-sich-von-einem-roboter-operieren-lassen-a-600562 (top left) | https://www.theguardian.com/technology/2015/feb/18/artificial-intelligence-nanotechnology-risks-human-civilisation (top right) | https://www.handelsblatt.com/unternehmen/beruf-und-buero/buero-special/kuenstliche-intelligenz-die-robo-anwaelte-kommen/13601888.html (bottom left) | https://www.faz.net/aktuell/feuilleton/medien/medien-chatbots-roboter-die-nachrichten-ansagen-14363427.html (bottom right). Screenshots taken on 2019-12-06.

2017). Here, next to the same popular pictures of humanoid robots, we can find a variety of "awful stock photos" (Geitgey, 2018) showing "AI tropes" like wires connected to a brain, or a "human face coalescing … from the atomic parts of the AI" (Winfield, 2017) – similarly fostering the impression that artificial intelligence has human characteristics.

Pictures of humanoids are the most visible and obvious way in which animacy is attributed to robots in the context of media discourse. The humanoid bias goes beyond ascribing just humanoid embodiment to robots, however, and extends into to other forms anthropomorphism. While only few articles state outright that a certain robot "is basically a human" (Reed, 2016), robots and neighboring technologies are frequently framed as having human-like features even beyond their physical shape.

Indirectly, the choice of specific topics covered in the media sets the basis for this practice. Sarah Kriz and colleagues (2010) found that, while the recent boom in social robotics was covered extensively in the media, advancements in robots' cognitive capabilities were systematically given less attention. Some technology news sites even set the whole focus of their robotics sections on social aspects of robotics. Robots are frequently framed as being about to take over not only specific tasks traditionally performed by humans, but complete jobs or even roles. Examples range from specifically social roles ("Zora, the Robot Caregiver", Satariano, Peltier, & Kostyukov, 2018; "transforms into a cute robot companion", Gibbs, 2016a), to service tasks ("turns into an adorable mini robot butler", ibid.; "newest crew member of Costa Cruise Line", Reese, 2016) and white collar jobs ("the new bookeeper is a robot", Monga, 2015; "the robo-lawyers are coming", Postinett, 2016), to law enforcement ("humanoid robots invade our lives as ... first responders", Belfiore, 2014) and even management ("why a robot could be the best boss you've ever had", Chamorro-Premuzic, 2016). Especially on technology news portals and articles in the science-oriented mode, readers are frequently invited to "meet" a robot or read that humans "welcome" a new robot: "Meet Luigi the PoopBot. He's Here to Scrape Your Sewers" (Grey Ellis, 2016) "Meet Zora, the Robot Caregiver" (Satariano et al., 2018), "[A baseball manager] is Welcoming Robot Umpires" (Phillips, 2019), "People will welcome the new [robot] 'master chef' to their kitchen" (Joshi, 2018). These phrases not only attribute personhood to a robot, they also frame it as being able to interactive socially. Moreover, it depicts robots as individuals – even though they usually are not even unique one-off artifacts, but in fact off-the-shelf robot models produced by the hundreds and thousands. Frequently, these depictions do not explicitly refer to human characteristics but more generally to characteristics of living beings. We find references to robots with not only a humanoid form, but with biological bodies and physiological processes – for example, when a sensor technology for the detection of heat or pressure is called an "electronic skin that could allow robots to 'feel' pain" (Boland, 2019).

Emotions are often attributed to robot technology as well. A demonstration video by Boston Dynamics (2016), showcasing their Atlas robot being repeatedly pushed in order to demonstrate its balance and stability, caused considerable media reactions attributing suffering to the stumbling, "abused" robot, even describing the actions shown in the video as "bullying" or "torture", and predicting the robot's future revenge (e.g. Hern, 2016; Koerber, 2016; Novak, 2016; Stockton, 2016).

When describing so-called "intelligent" technologies, field experts themselves frequently employ terms loaded with meaning from a human perspective, such as "experience", "learning", "recognizing", or "thinking". Not surprisingly, the news media pick up these terms and relay them to their non-expert audience – which is not able to assess subtleties like the difference between machine intelligence and human intelligence. Headlines like "Facebook is Training Robots to Think" (Paris, 2019) or "Robotic Vacuum Remembers Your Home's Layout" (Verger, 2018) thus insinuate that robots have human-like cognitive processes.

Voices from within the robotics and AI community have been criticizing the liberal and sometimes incautious use of these "suitcase words" (Minsky, 2006, p. 11). Robotics professor Rodney Brooks (2017b) noted that "the use of these words suggests that there is much more there than is there" and expressed worries that "people will over generalize and think that machines are on the very door step of human-like capabilities in these aspects of being intelligent". Section 5.6 will dive deeper into this critical discourse.

Be it physical shape, physiological or cognitive processes, professional and social roles, or even personhood and individuality – in media coverage of robot technology we can observe references and comparisons to human characteristics on all levels. Through science fiction references or comparisons to real humans, robots and other autonomous technology is made tangible, imaginable, and exciting for the non-expert audience. However, not only the characteristics of specific technologies are at the center of the discourse. The consequences of their deployment, too, are discussed in a way that inherently frames robot technology as quasi-animate.

The Inevitable Robot Apocalypse: Commenting on a Technologized Society

When exploring media discourse and looking specifically for pictures or mentions of robot technology, the majority of articles one encounters does not focus on specific technologies. Instead, we find a steady stream of articles on automation and its consequences that not only make constant references to robot technology but also embed it in a very specific narrative: That of robots actively and intentionally approaching to steal human jobs. This narrative – in its extremes packed with references to "robot overlords" and even the end of humanity – is not only "a convenient, faintly ominous, and click-generating shorthand for referring to the phenomenon of automation in the workplace" (Merchant, 2019). It is also a way to express the perceived inevitability of ever-increasing automation, and the apparent helplessness of the average person in the face of more and more traditionally human tasks being automated. This process goes beyond the automation of simple physical tasks and increasingly includes "soft" skills like social interaction, caretaking, or creativity. In the context of media discourse, the perceived inevitability of advancing autonomous technology is – once again – met with references to a curiously specific science fiction-inspired narrative. Not only is the active involvement of human agents pushed to the background. Robots – here understood as a stand-in for automation in general – are framed to be coming from a predetermined future. These robots want to steal human jobs, and a takeover of robot overlords appears to be practically inevitable (e.g. Corbyn, 2015).

Presumably, most journalists and readers are well aware that it is not robots who decide to purchase and install themselves in factories and personally fire the employees whose tasks they take over. As Astra Taylor (2018) suggests, a brutally honest headline would be: "Capitalists are making targeted investments in robots designed to weaken and replace human workers so they can get even richer". Obviously this is not very catchy. Therefore, typical headlines instead refer to an active, physical approach of robots – such as: "Robots are Leaving the Factory Floor and Heading for Your Desk – and Your Job" (Corbyn, 2015). Robots are almost routinely framed as acting on their own initiative, while human agents – such as manufacturers or customers – are effectively backgrounded (Leeuwen, 2008, p. 29). An analysis of the representation of military robots in the mass media and in Department of Defense press releases similarly observed that, while there usually "is some reference to the social actor elsewhere in the text, ... their role in the represented action has been de-emphasized" (Roderick, 2010, p. 238).

When robots are described as coming for human jobs, there is already a certain attribution of intentionality as the driving force: "Yes, the Machines are Getting Smarter, and They're Coming for More and More Jobs" (Tufekci, 2015). This intentionality is frequently framed as something dangerous, the idea being that once we let robots out of our control they will follow their own – often malicious – agenda. Individual reports and articles usually only use one or two specific references. Taken together, these references assemble into a curiously stable narrative that sounds as if plucked directly from a science fiction novel. A science fiction novel with a decades-old and very specific niche narrative in which robots' agency and intentionality are directed against humanity. A narrative that reportedly already felt outdated to science fiction legend Isaac Asimov himself in the 1950s (Nof, 1985, p. xi). This narrative of robots rebelling against their human masters is a common trope in science fiction literature of the twentieth and twenty-first century (cf. Chapter 1, Section 1.3). It mirrors the historically troubled and competitive relationship between workers and machines, as well as the perceived threat to human agency over the means of production (Meinecke & Voss, 2018). This narrative can be traced back to the Czech play "R.U.R." (Čapek, 1920) and has been retold and modified many times, from short stories by Isaac Asimov to television series like "Battlestar Galactica" (Moore, 2004) and movies like "Ex Machina" (Garland, 2014). Inspired by this, autonomous technology is described in the news media as, for example, being "on a mission" (Alba, 2016), "eager" to do something (Glaser, 2016), "willing to kill" (Sample, 2016), needing to be "tamed" (Stone, 2015), being allowed to "have free reign of the house" (Gibbs, 2016b), "stealing" jobs (Vardi, 2016), "infiltrating ... assembly line[s]" (Paur, 2013) or "escap[ing] the factory floor and star[ting] conquering big cities" (Simon, 2017a). It is no accident that the plot line of the Terminator movie franchise – in which machines take over control of the world – is one of the most referenced stories.

The narrative of robots' negative power over humanity moves between stories of a more or less unfriendly competition on the labor market and predictions of outright genocide. Next to many articles framing robots as only interested in our jobs there is a huge section of the discourse discussing robots as a threat to human identity (cf. Złotowski, Yogeeswaran, & Bartneck, 2017), featuring very explicit descriptions of an apocalyptic future in which robots will destroy humanity. References range from a robot vacuum "trying to eat its owner's head" (McHugh, 2015), to smart robots "casting out workers from factories and offices" (Hagelüken, 2016), to "AI, robotics, and autonomous vehicles all unit[ing] in a winner-takes-all battle against humanity itself" (Bishop, 2014). The basic plot of robots wanting to take our jobs is followed by the idea that robots will "go rogue" at some point (e.g. Ambasna-Jones, 2016), breaking loose from human control, even plotting revenge against the humans who treated them badly. This specific narrative was particularly popular in the coverage of the "tortured" Atlas robot: "Engineers have been filmed beating, pushing and torturing a humanoid robot in video that could one day be seen as the beginning of the war between man and machine" (Griffin, 2016), "When robots inevitably take over the world, remember this video" (Koerber, 2016). The narrative continues with robots actively trying to take control over the world in a "rise", "march", "uprising", "rebellion", or "revolution", in which they are expected to resort to outright, systematic violence. For example, Amazon's warehouse robots are frequently referred to as an "army" (e.g. Chang, 2014; Thielman, 2016) and robots posing a potential competition for human workers are often framed as an "attack", "invasion", or "war" (e.g. Epstein, 2016). Finally, the narrative goes, robots will "take over" (e.g. Bostedt, 2016) and assume the role of overlords.[7]

In some variations, the narrative even escalates towards an apocalyptic end of humanity. In line with the idea that eventually all traditionally human jobs will be automated, this dystopian narrative frames the conflict with robots as something not only relevant for those whose jobs are threatened, but for all of humanity. Robots becoming more and more similar to humans triggers, on the one hand, discussions of human-machine distinction, of "what it means to be human" (Chatfield, 2016) and the "future of (hu)mankind" (Burton-Hill, 2016; McMahon, 2015). On the other hand, it causes robots to be perceived as a threat to our very humanness – be it as competitors in a "race" (Hagelüken, 2016; Thielman, 2015), opponents in "mankind's war with the robots" (Epstein, 2016), or as an outright existential threat of "the end of humanity" (Biggs, 2016; Robbins, 2016). This highly dramatic timeline is a conglomerate of tropes, but almost never referred to in its entirety. Instead, news articles usually refer to little snippets of the narrative, implying that readers already know the story very well. So well, that important aspects of the narrative are often used with the definite article, and occasionally even capitalized like proper nouns – such as "The Robot Revolution" (instead of "a robot revolution").

7 The trope of robotic overlords has become so common that a mocking counter-trope has appeared. It can be traced back to the quote "I, for one, welcome our new insect overlords" from the film adaptation (B. I. Gordon, 1977) of a H.G. Wells (1905) short story. "Its phrasal template 'I, for one, welcome our new X overlords' has been widely used to express mock submission towards an obsessively controlling individual for the sake of humor" (Know Your Meme, 2012). The variation "I, for one, welcome our new robot overlords" has become such a staple that it is frequently used for introductions of new robot technology.

At the same time, this specific narrative is presented not only as a predetermined future waiting for the present to catch up, but also as a physical location from which certain robots are coming from. Here, robots are typically described in a way that attributes a strong agency and even intentionality to them. They are approaching us from the future, their arrival apparently something we as humans do not have any control over, with all agency lying in the hands of the robots themselves: "2017 was the Year the Robots Really, Truly Arrived" (Simon, 2017b).

The highly dramatic narrative chain of events is not only referenced over and over, it is persistently framed as not as a possibility, as a hypothetical, fictional future, but as a predetermined timeline of future robot technology development. This way, articles reporting on technologies of the present also appear to prepare the audience for a "known" future in which robots will take over: It is "only a matter of time" (Hamill, 2016), "the only question is when, not if, humanoid robots will work, play and war beside us" (Belfiore, 2014). The present simply appears as not having reached the start of this timeline yet. That it will start at some point is undoubted: "The day ... draws inexorably closer" (ibid.). Even present technological developments are constantly compared to their "predicted" counterparts. Whenever a sufficient overlap is perceived it is interpreted as the future having "arrived" (Fetterman, 2016) or us "already living in the future" (Stone, 2015). Certain developments in robotics are even referred to as a sign that the anticipated disastrous ending – the Robot Apocalypse – is already in sight. Robot platforms measuring up to the expectations set by fiction are therefore popular illustrations for anything considered remotely robotic – such as the few commercially available humanoids (e.g. Pepper, Nao) and those staged as sufficiently impressive in demonstration videos (e.g. Atlas). If, however, a new robot technology does not live up to the predictions and expectations, this is not necessarily understood as indication for the predictions being wrong. Instead, it is described as a temporary delay – typically observable in comments that some expected robot capability is "not yet there". For example, in the 2015 DARPA robotics challenge, research teams competed by having their state of the art robots complete an obstacle course simulating the aftermath of a disaster like the Fukushima catastrophe. From a robot technology perspective, the – mostly humanoid – robots showed impressive abilities. Media reactions, however, were steeped in both apparent Schadenfreude and relief in the view of robots failing to complete seemingly simple tasks like opening a door (Guizzo & Ackerman, 2015). Even renowned robotics professor Rodney Brooks was compelled to comment that "anyone who is worried about the robot apocalypse just needs to keep their doors closed" (cited in Keay, 2015). However, even in the face of the slapstick-like robot fails the discourse was steered towards the apparently inevitable Robot Takeover. Readers were asked to "laugh ... while [they] can" (Tabarrok, 2015), "with whatever short-lived impunity [they] may still have" (O'Connor, 2015).

Even when the alluded-to future is not one explicitly predicted to end in the Robot Apocalypse, robot technology is almost always expected to evolve into something more human-like than today. Even the robotics news site Robohub, analyzing past and possible future trends in robots in an article titled "Envisioning the Future of Robotics", illustrated the predicted evolution of robots from less to more humanoid, referencing famous illustrations of apes "evolving" into humans (Mayoral Vilches, 2017). In this sense, attributions of animacy to robots are used to show how close to the future a specific technology is perceived to be.

5.5. Switching Perspectives: In/Animacy Attributions as Constructive Practice

The role of animacy attributions appears to be minor when they are quantified by the proportion of text they take up in individual articles. However, the typical positioning of animacy attributions within and around an article gives them substantial impact. In the context of specific news articles, "flashy" references to robot animacy are usually positioned strategically where they can attract attention or drive home a critical standpoint on automation. Typically, these references can be found in a headline, "punchline", or illustration. In this, they often stand in contrast or even contradiction to the tone and content of the rest of an article, which typically discusses robot technology in a more matter-of-fact manner. A headline like "Robots Instigate Revolution in the Workshop"[8] (Menzel, 2017) conjures images of humanoid robots mounting the barricades in a factory hall. The article below this headline, however, simply reports that "car manufacturers will increase the use of robots" and does not imply in any way that robots are deciding anything by themselves – least of all a revolution. Under the headline "Plant Biologists Welcome Their

8 Translated from German by the author.

Robot Overlords" – raising visions of similarly science-fiction inspired scenarios – we read that "old-school areas of plant biology are getting tech upgrades that herald more detailed, faster data collection" and no mentions of overlords (Ledford, 2017). And under the headline "New York State Creates Group to Study Rise of Robots" we do not find a report on scientists observing the approach of a robot army but learn that "New York state will convene a panel to study the impact of artificial intelligence, robotics and automation on the state and suggest areas of potential regulation" (Vielkind 2019). And vice versa: A concluding sentence like "The idea is that biohybrids could be very cheap to produce, and they would just biodegrade once they outlive their usefulness" is followed by a snarky "Or, you know, they could start multiplying and take over the world" (C. Smith, 2016).

Articles discussing an emerging robot technology expected to "inevitably" (cf. Section 5.4) play an important role in the future often use a combination of the negation of animacy and the word "yet". A sentence or headline might explain that robots do not have certain advanced abilities, which could make them appear animate. Crucially, however, the sentence ends with "yet" – implying that it is only a matter of time until those abilities emerge: "Robots aren't stealing our jobs, yet" (O. Smith, 2015).

In the in the vast majority of cases we find animacy attributions not in the text itself but in the image(s) placed next to the text. Especially pictures of humanoid robots – both real and fictional ones – are ubiquitous (cf. Section 5.4) and often create an almost absurd contrast to the written article. For example, an article which explicitly tries to distance itself from science-fiction tropes by stating "We are not talking about the artificial intelligence robots of Hollywood dreamers ... In the real world this is much more mundane and more immediate", is accompanied by a picture of a fictional humanoid robot from the movie "Robot & Frank" (Schreier, 2012), captioned "Could fiction soon become reality?" (Ambasna-Jones, 2015). An article on autonomous weapons, referred to as "killer robots" in the headline, is illustrated with a picture of small humanoid Nao robots playing soccer (Dreifus, 2019). An article on image recognition software is placed next to a big picture of the humanoid Pepper (Schmundt, 2019).

While interacting with such an article readers are submitted to a constant switching of perspectives on robot technology: From a headline alluding to robots having intentionality ("Want to steal your job"), to a dry report on the management of a company announcing to automate certain production tasks, to a punchline referencing robot overlords, to an illustration featuring a cute humanoid robot, captioned with a neutral reference to the article text. Over the course of the reading process the technology in question changes frames, from "animate" to "inanimate" and back, several times. While tone and focus vary with the type of article and publication, this switching is observable across the whole sample of analyzed articles, independent of the style.

We encountered this constant change of perspective already at our earlier stops along the life cycle of robots, in our explorations of research and development practices, of demonstrations, science communication, and marketing. Just as in these other contexts, also in media discourse the switching of attributions of in/animacy to robots has context-specific constructive functions. In media coverage of both, specific robotic technologies and technological progress in general, pictures of futuristic humanoid robots and references to dramatic dystopian narratives appear to do an excellent job at attracting readers' attention. Not only that, comparisons to well-known fictional narratives can make a difficult to grasp technology tangible and imaginable. Moreover, references to well-known tropes like the Robot Apocalypse can serve as a comment on the seemingly inevitable saturation of society with autonomous technologies.

At the same time, the majority of analyzed articles also attempts to give a reality-based perspective on the technologies in question, focusing on technical details and limitations. The constant switching of perspectives, between robot animacy and inanimacy, has a constructive quality. It is a reflection of the complexity of the topic, and of journalists' attempts at satisfying multiple demands: That of attracting an audience, that of informing the audience, and that of commenting on current societal developments.

However, it appears that attributions to robot animacy, the forms discussed above, have become such a staple in media discourse that their use has become somewhat opportunistic. References to science fiction and images of humanoid robots often are not selectively and constructively sprinkled among fact-based information, but appear to be routinely pasted on everything that remotely resembles a robot. This practice has been drawing criticism, as the next section will discuss in depth.

5.6. Critical Discourse: Animacy Attributions as Traffic Bait?

The common practice of overpowering depictions of robots as inanimate machines with Terminator pictures and references to the Robot Apocalypse does not only generate favorable reactions. The treatment of fictional narratives as predictions of the future, painting a picture of an inevitable dystopian future of humanoid robot overlords, as well as the frequent use of misleading headlines and visualizations – a practice which is usually considered a journalistic faux-pas (Schenk, 2007, p. 122) – are facing criticism from both within the journalism community and the robotics and AI community.

With the mass media as a main source of information on scientific developments for the lay public and policy makers (Schenk, 2001; Summ & Volpers, 2016), the way robot technology is presented and framed in the media can have far reaching consequences – from "setting the agenda" (Lippmann, 1922; McCombs & Shaw, 1972) to critically shaping public opinion, legislation, and further technological progress (as Chapter 7 will discuss in more depth). For example, the New York Time's coverage of robotics promoted the extensive development and use of social robotics in the US (Russett, 2011) – not by directly praising certain robot technologies, but "by creating … opportunities" or "niches" which robotics "could usefully occupy" (Arthur, 2010, p. 174).

Interestingly, the media-critical discourse also takes place within the same publications that regularly employ the very practices facing criticism. The Guardian – which routinely features articles referencing robot animacy and illustrates most of its articles on AI with pictures of humanoid robots – also published articles titled "The Media are Unwittingly Selling Us an AI Fantasy" (Naughton, 2019) and "How the Media Gets AI Alarmingly Wrong" (Schwartz, 2018).

While academic publications on the issue are still rare, the bulk of critical reactions ranges from heated discussions on social media among AI ethicists and communication scientists to dedicated journalistic articles. They point out the "epidemic of AI misinformation" (Marcus, 2019b), the "fantasy-based" (Fernaeus et al., 2009, p. 280) "unhinged discourse" (Schwartz, 2018), and that "robots aren't going to kill you" (Buchanan, 2015). Some of the criticism is quite harsh: Journalists are accused of being "clueless", "willfully ignorant" (Sofge, 2015) and "opportunistic" (Schwartz, 2018) when it comes to robots and AI, "callously traffic-baiting" (Sofge, 2015), "spreading misconceptions" (A. Guzman, 2017), "misrepresenting research for the purpose of generating retweets and clicks" (Schwartz, 2018), and "amplifying industry's self-interested claims" (Naughton, 2019).

Especially the ubiquitous references to science fiction narratives face criticism. For example, Isabella Hermann (2019a) argues that they are not a suitable base for a societal discourse on AI; that, on the contrary, they distract from the really relevant opportunities and challenges. Hermann (2019b) also points out that "robots in films ... tell us little about technical progress or the pressing challenges of digitalization and artificial intelligence, but all the more about ourselves". Similarly, Lisa Meinecke and I (2018, p. 208) argued that considering science fiction as a kind of societal wish list for the future means disregarding that these narratives are not a neutral repository of ideas about technology or a road map to the future. Rather, they are a reflection of the values, hopes, and anxieties of the cultural context they originate from.

This criticism of the animistic treatment of robot technology is particularly directed at the media. However, as we saw in Chapters 3 and 4, also roboticists themselves face this criticism – be it in the context of robot design, human-robot-interaction design, or science communication and marketing. We will revisit this issue in the following final chapter.

5.7. Summary

References to robot animacy are ubiquitous in the news media: In discourses of promise or peril, of fascination or repulsion, and across all journalistic styles; in science-oriented and problem-oriented articles, in enthusiastic reports on a new robot technology, in pessimistic essays lamenting the imminence of a machine-controlled dystopia, and between dry descriptions of the newest sales figures of industrial robotics in the automotive industry. In all these contexts, robots and other autonomous technology are regularly framed and (re)presented as possessing human-shaped, even biological, bodies, as well as emotions, goals, and intentions.

Most instances of animacy attribution in media discourse can be observed in one of two contexts. Firstly, in the context of discussions of autonomous and autonomous-appearing technology – independently of whether this technology is really "robotic" in the technical sense. References to human-like characteristics are used to make complex and difficult to grasp technologies tangible and imaginable for a lay audience. In this context, animacy attributions range from comparisons with real and fictional humanoids to references to traditionally human tasks and even mentions of personhood and individuality. Secondly, we can observe animacy attributions in commentaries on the seemingly inevitable saturation of our environment with autonomous(-appearing) technologies. In this context, we find references to an apparently "predetermined" future – strongly inspired by specific science fiction narratives in which malicious robots want to, or even succeed to, “take over” the world.

While most instances of animacy attribution take up little space, their typically opportunistic placement in headlines, punchlines, and pictures gives them considerable force, letting them overpower any otherwise fact-focused depictions of robot technology. The impression while reading such an article is that of a constant switching of perspectives, from robots as animate beings, to robots as inanimate machines, and back again. This switching reflects the multiple challenges of reporting on a complex emerging technology: attracting the audience's attention, informing on the technology, making it tangible and imaginable, and at the same time commenting on its societal consequences.

This predominant style of media reporting on robot technology and artificial intelligence is at the center of a lively critical discourse – both in the communication science and in the robotics and AI community. Critics are concerned that the framing of robot technology as quasi-animate constitutes a dangerous misrepresentation of the current state of the art of robot and AI technology.

6. Conclusions ... and Openings

6.1. A Recapitulation

We have come a long way on our tour of the life cycle of robots, since our first step into the current New Age of Robotics. Before recapitulating the insights we gained along the way, let us jump back to the beginning and return to mind why we went on this journey.

We started off with the realization that robots, as well as their technological and fictional ancestors, have been a part of our world for a long time, and that today they are more in the focus of attention than ever. Robots are at the center of a hype and the subject of a controversial and emotional discourse across political, economic, academic, and other public spheres. On the one hand, robot technology is hailed as a utopian panacea, as a solution for a range of fundamental problems our society faces. On the other hand, robots are also inextricably embedded in a dystopian, at times even apocalyptical, narrative of humans losing control, of robots in competition to us, even of our very humanness being in peril.

This is rooted in the fact that robots, although they are new in our direct physical environment, are not new to us at all: Across both extremes of the discourse, technological reality is inherently intermixed with decade-, even century-old, fictional narratives. Even as robot technology is just now starting to step out of its factory cages – into our everyday lives and direct physical environment, our jobs, our households – the realm where robots are present most vividly is still our imagination. The rich cultural history of literary and cinematic robot fiction fundamentally influences our relationship with the robot technology in our real life – where the so-called New Age of Robotics brings forth technologies that appear to move closer and closer to what we know from fictional narratives. These new technological artifacts are not only used in physical proximity to us but are also embodied and mobile, act autonomously, and sense and manipulate their environment. Some are even socially interactive and have a human-like design.

In this newly close relationship with robot technology, we can observe instances of humans perceiving them and treating them like more than "just" inanimate objects. Are they to be considered machines? Creatures? Something in between? Something else entirely? It seems that robots do not belong in the ontological category of the "inanimate object". But apparently they do not belong with the "living beings" either. In this, robot technology has been redrawing the attention of academia to the phenomenon of attributions of animacy to technological artifacts.

Humans have always attributed characteristics of living beings to inanimate objects – a human quirk that, historically, has often been viewed as either "primitive" behavior or, in the context of science, as methodological misconduct. With the advent of the "New Age of Robotics", however, a new type of inanimate object has been pushing into humans' lives: machines with features traditionally reserved for animate beings, technological artifacts appearing to sway between the natural and the artificial (cf. Haraway, 1991, p. 152).

This unique ontological challenge posed by robots and other autonomous technologies inspired the overarching research question of this book. Crucially, this question is not which ontological category robots "really" belong to. Rather, we followed Lucy Suchman's suggestion to shift the discussion from "whether humans and machines are the same or different to how and when the categories of human or machine become relevant, how relations of sameness or difference between them are enacted on particular occasions, and with what discursive and material consequences" (2007, p. 2).

Inspired by this, we explored the range of discursive and non-discursive manifestations of in/animacy attributions to robotics, as well as their conditions, functions, and consequences. We made four major observations:

First, attributions of animacy to robots are not an isolated phenomenon, not just a perceptual quirk of the human mind, and they are not only present in the direct, physical interaction of humans and robots. On the contrary, we found a broad range of manifestations, on a discursive, non-discursive and material level along the whole life cycle of robots: Visions of a future shared with socially interactive robot companions – both in fictional narratives and in the sociotechnical futures, which are present in the roadmaps and guidelines of the robotics industry, and of academic and policy institutions (Chapter 1). Practices of making robots "lifelike" in human-robot-interaction research

(Chapter 2). Robots being perceived as unpredictable but valuable team members in robotics research and development (Chapter 3). Complex narrative scenarios and robot personas staged for science communication, marketing, and demonstration practices (Chapter 4). And constant references to fictional narratives of robots as competition to humans in media discourse (Chapter 5).

Second, attributions of animacy to robots are not a static, inflexible practice. Instead, across all the explored contexts we found a constant switching of perspectives, of robots being perceived and represented alternatingly as inanimate objects and animate beings.

Third, these attributions of in/animacy to robots – and the switching between them – are not just an involuntary reaction to certain features of the robot, such as its design or its behavior, but in fact have context specific constructive qualities. Robot technology is embedded not only in a rich fictional and cultural history but also in a quite controversial public discourse. In this context, attributions of in/animacy help us grasp and embrace the unique challenges that robot technology brings to our lives. They are powerful in that they shape our perception of, and our relationship with intelligent technologies. In this sense, they can be a way of navigating the complex environment of our technologized society.

Fourth, the ubiquitous attributions of in/animacy to robots are persistently accompanied by a critical discourse. This criticism is predominantly aimed at the deceptive, overly opportunistic, or unthinking use of these attributions, which are feared to cause systemic, lasting, and potentially problematic consequences on policy, governance, and legislative decisions. The present chapter will discuss this in more depth (Section 6.3).

But first, Section 6.2 will take a step back to reflect on these major observations from a more cross-contextual perspective, identifying the underlying conditions, motives, and forms of in/animacy attributions permeating the whole life cycle of robots.

Section 6.4 will step even further back and break out of the robot life cycle. We will have look around in the vicinity to see how practices of in/animacy attribution and the related critical discourse, as well as the more general question of what is "natural" and what is "artificial", are discussed in our technologized society.

Finally, in Section 6.5, I will step down from the observer position and make some suggestions for how the insights of this book can be applied constructively by different actors.

6.2. The Constructive Quality of In/Animacy Attributions

Across the different contexts of robotics, we encountered attributions of animacy to robots in a variety of discursive, performative, and material forms. From playful to opportunistic, from reflective to unthinking; from images of humanoids and other comparisons with the human physiology, to references to science fiction stories in which robots have life-like characteristics, to the creation of narrative robot personas, down to the linguistic level, for example in the use of gendered pronouns. Constantly switching perspectives, to and from attributions of inanimacy, were observable on different contextual levels: In the location of presentation (website vs. social media; headline vs. article body; physical work practice vs. academic publications), in the narrative perspective (robot vs. human as narrative persona), and in the task or goal specific to the context (e.g. educating and informing vs. attracting attention). Through the whole spectrum of contexts, across the various manifestations of attribution and the different levels of perspective changes, there was one aspect we found again and again: the underlying constructive function of in/animacy attributions.

Embracing Robots' Agency

Chapter 3 found that, in the particular context of robotics research and development (R&D), it is part of a robot's job to act unpredictably. Especially in the "troubleshooting" phases of robotics R&D, when the complex system of the robot clashes with the complex physical environment, roboticists are dependent on the robot as a "feedback-giver" – the robot's behavior serving as a crucial source of information. As we observed, this can lead to a perceived distribution of agency between the roboticist and the robot – reflected in roboticists perceiving their robots as something akin to a research companion or team member. This perception is reinforced by the often very heterogeneous and interdisciplinary structure of roboticist teams: The robot platform serves as a central object of focus, taking the role of a boundary object. In this context, attributions of animacy are a reflection of the robot's central and active role. At the same time, however, roboticists are required to take a strictly "professional" (i.e. technical) perspective on their robots, focusing on them as the inanimate machines they are, and refraining from openly expressing attributions of animacy – for example when writing research papers or technical documentation, or when presenting robot demonstrators to academic peers. The practice of constantly switching perspectives – from the robot as an animate research companion, to the robot as an inanimate object – is a constructive way of dealing with the multiple, constantly changing demands of a roboticist's professional environment.

Making Robot Behavior Explainable

In the context of robotics research and development practice (Chapter 3) we also encountered a phenomenon that earlier research in HRI and communication studies also found for the direct interaction of users and robots (cf. Chapter 2, Section 2.1): Observing a robot acting unpredictably appears to make humans think and talk about it as if it were alive. Crucially, this is not only the case for lay users, for whom the complex technical processes responsible for the robot's behavior are hidden inside a "black box". Also roboticists with professional insight and understanding of the technical details are prone to think and talk about "their" robots as if they were animate – even though they are perfectly aware that the robots are in fact inanimate objects. In practice, this is reflected in roboticists' practices of assigning names and gender to robotic platforms, in the "joking" framing of technical components as body parts, of technical processes as physiological functions, and in the attribution of liveliness and personality to robots.

In this context, in/animacy attributions are constructive in that they are a way of dealing with the strangely contradictory situation of an inanimate object acting in a way usually unique to animate beings. Allowing oneself to think and speak about a robot as a quasi-animate being makes its behavior appear somewhat more explainable (cf. Frey & Jonas, 2002). Crucially, this animate perspective is not static. Rather, human users and observers of robots appear to be able to effortlessly, playfully balance attributions of animacy and inanimacy. This allows them to keep a flexible perspective that is able to do justice to the robot's unique behavioral characteristics and its category-defying ontological status.

Making Robots Tangible

Another function of in/animacy attributions we encountered in several of the explored contexts is that of making robot technology tangible. Robots are a complex technology and, for most people, not yet a part of everyday life. This can make it difficult to imagine what the application of robot technology in our direct environment could look and feel like. Luckily, for those trying to bring across the vision of a life with robots, while real robots are still quite exotic for most, there is a rich cultural reservoir of popular science fiction narratives to draw from. By referencing well-known fictional robot characters, science communicators and media professionals try to make robot technology imaginable and tangible for their audience (cf. Chapters 4 and 5). These fictional narratives are so popular and engaging because they deal with the topics of robotics in relation to ourselves. Classic narratives center on robot characters in juxtaposition to us "real" humans, on robots as the "other", as potential companions or competition, and on their struggle of "wanting to be like us". It is these narratives that media discourse draws from in an effort to make complex topics – such as the consequences of increasing automation on our lives – tangible to the audience. For a lay audience it is simply easier to visualize an army of humanoid robot overlords physically coming to steal their jobs from under their noses, than to disentangle difficult to grasp concepts such as "artificial intelligence", "machine learning" or "algorithms", and the systemic effects they will have on the labor market.

Even when communication efforts do not explicitly refer to science fiction narratives, comparisons with the human body, with its physiology, its sensory and emotional experiences, are ubiquitous. We find an abundance of illustrations of humanoid and human-like robots all over robotics innovation roadmaps, policy documents, research institutions' social media accounts, and newspaper articles – regardless of whether the robot technology in question is actually humanoid, or even strictly speaking a robot. Importantly, it is not the goal of these communication efforts to convince the audience that the robot technology in question is actually animate or human-like. In fact, most of the communication activities also present quite a "technical" perspective, focusing on the robot as a – clearly inanimate – machine. Wherever it fits the purpose, however, wherever the technology needs to be made tangible for the audience, we encounter a switch to attributions of animacy – and back to inanimacy. This does not only include references to the physical shape of the human body, such as illustrations of complete humanoid robots. We also find cartoons adding cute eyes to a space probe, descriptions of robots as a "he" or "she", and even complete social media accounts from the first-person perspective of a robot persona, reporting on exiting adventures, sensory experiences, and social interactions with "friends", "family", and "colleagues".

Making Robots Desirable

Across several contexts along the life cycle of robotics, we were able to observe that attributions of animacy to robots are utilized to make robots desirable – figuratively and literally.

Robots are made desirable in the figurative sense, for example in the contexts of science communication, marketing, and media discourse (cf. Chapters 4 and 5). Here, instances of animacy attribution serve to attract the attention of the audience – of potential customers, investors, and sponsors, but also of the broader public. Actors in these contexts therefore present robots in exciting interactive scenarios, embed them in narratives referring to desires and struggles relatable to the audience, in stories known from popular culture. Likewise, a headline referring to the robot revolution or a picture featuring the Terminator simply draw more attention, more readers, more clicks, to a news article than a headline or picture depicting robot technology as a "boring" and difficult to understand technical object (cf. Chapter 5).

Attributions of animacy are also employed to make robots desirable in a more literal sense. Staging robots with a quirky personality and an exciting life story, letting them have funny and adorable interactions, even letting them speak for themselves as a persona in the first person perspective, makes them engaging and likeable. In the context of science communication (cf. Chapter 4), this is utilized to draw positive attention to both robotics research and development, and to projects using robot technology to achieve scientific goals. We observed personas like Roboy, who takes his audience along to events and invites students to "hack him". We followed the travels of spacecraft and planetary rovers, who explore the universe, interact with their team of human engineers, and share information on their newest discoveries. Crucially, these narratives of animacy are "switched on and off" whenever it appears useful. In some cases, a robot's first-person-perspective Twitter account is accompanied by a website on which the robot's technical details are described with scientific distance. In other cases, social media posts "by the robot itself" take turns with posts by the engineers behind the project. Always, the goal is to convince the audience that the work done in the context of the respective projects is interesting, successful, and worth the taxes the audience might have contributed to the efforts.

In the context of commercial marketing (cf. Chapter 4) we found that attributions of animacy are switched on and off in an even more opportunistic way. Here, the goal is quite obviously to sell robots as a product. When iRobot presents its Roomba vacuum cleaners as dutiful, dedicated, pet-like cleaning companions, it does so in order to make potential customers wish to own one. Here as well, attributions of animacy are only one half of the effort: One click further, on the company's website, the robots are presented as thoroughly inanimate products for sale, their technical details in the focus of the presentations, instead of their cute personality.

Chapter 2 (Section 2.1) discussed how human-robot interaction research (HRI) approaches the issue of animacy attributions to robots. In taking this phenomenon seriously, and in striving to use it in a constructive way, the field of HRI studies is progressing away from the historically mistrusting perspective on the phenomenon (cf. Section 2.3). Again, we were able to observe that animacy attributions to robots are viewed as having a certain function and are researched and employed with an overarching goal in mind. We can find a large number of HRI studies trying to identify and quantify features of robots (e.g. design, behavior) and users (e.g. personality, cultural background) that reliably trigger animacy attributions. Drawing heavily on research in the cognitive sciences, these studies strive to make animacy attributions controllable and predictable. The goal of these research efforts is usually to facilitate or optimize human-robot interaction and, in the long-term, to promote the introduction to interactive robot technology in society at large.

Making Robots Imaginable

Another function of animacy attributions we were able to observe is that of making robots imaginable. This can be understood on two levels: On the one hand, in the sense of making robot function imaginable for the future use of the robot; and on the other hand, in the sense of making robot technology imaginable for a future society.

In robotics demonstrations (cf. Chapter 4), be it in academic or commercial contexts, animacy attributions are employed purposefully to "prove" that a robot is functioning as claimed, now and in the future, with as little human intervention as possible. We observed that demonstrations are therefore carefully scripted, rehearsed, and – in the case of video demonstrations – also heavily edited performances. In order to make a robot appear as autonomous as possible, any "undesirable" human intervention is usually either subtly backgrounded or actively concealed. Additionally, some demonstrations embed their performance in a scenario inspired by fiction-inspired visions of a robotized future. The overall result is a performance in which the robot appears as autonomous as possible. Sometimes, a robot is even staged as more autonomous than it actually is, leaving the audience with the impression that the robot possesses something akin to animacy. Especially non-expert audiences cannot realistically assess the current state of technology. They can therefore easily misjudge a robot's performance and overestimate its autonomy – such as in the case of Boston Dynamics' videos, which regularly go viral and have viewers express the belief that a rise of robot overlords is imminent.

The goal of proving a robot's functionality, of making it tangible and desirable not only for the present moment but also for the future, is also observable in other contexts along the life cycle of robots. In interaction studies (cf. Chapter 2, Section 2.1), in demonstrations, and in science communication (cf. Chapter 4), robots are embedded in scenarios meant to illustrate desired futures. Countless robotics research and development projects present their results in scenarios in which robot technology is advanced enough to be deployed seamlessly in everyday life. Frequently, these scenarios make use of narratives of robot animacy. They paint a picture of smooth interaction and of companionship with the robot by subtly or blatantly referencing narratives that the audience is well acquainted with from science fiction. This makes the application of robots in desired futures not only imaginable, but also paints a picture of the robot's undisputable relevance, even necessity.

Not only in science communication and demonstrations, also in media discourse (cf. Chapter 5), attributions of animacy to robots are inherently connected to visions of and predictions for our technological future. In these contexts, the future we seem to "know" from science fiction – a future populated with agentic, intentionally acting human-like robots – is treated not simply as an entertaining story, but almost as a prediction. Especially in media discourse, references to robots as quasi-animate beings serve to paint a picture of robotic inevitability and are a way of commenting on the seemingly unstoppable advance of autonomous technologies in our everyday lives. Here, too, attributions of animacy are not used consistently, but switched on and off wherever they serve their purpose. This is the reason why we find so many fact-focused technical articles, the main text body focusing solely on the technology's clearly inanimate features, accompanied by pictures of humanoid robots, by flashy headlines and punchlines referencing robot overlords, the rise of the robots, the robot revolution, robots stealing jobs – over and over.

6.3. Critical Discourse: Individual and Systemic Issues

The way robot technology is often portrayed in the media – with its flood of Terminator pictures and constant references to an inevitable robot apocalypse – faces considerable criticism. In fact, many other practices of attributing animacy to robots are discussed controversially as well. On our tour along the life cycle of robots, we were able to identify similar points of criticism in almost all contexts. The following sections will revisit and consolidate them.

The critical discourse is directed, on the one hand, at attributions of animacy as potentially being problematic on the level of the individual. Here, the focus is on whether representations of robots as quasi-animate beings constitute a form of deception, in that they create misconceptions of current and unrealistic expectations for future robot technology. On the other hand, there is also a controversial discussion of animacy attributions having long-term consequences on a more systemic level, in that these misconceptions can influence political and legislative decisions – but also in that they draw away attention from equally, if not more important, social and ethical issues in the context robot technology.

The Individual Level

One overarching point of dispute is whether actively making robots appear animate – be it though their design, their behavior, or by setting them in a certain narrative frame – is a form of deception. Even when the intention is benign, such as when a robot's humanoid design is supposed to facilitate interaction, the question remains whether it is a harmless form of manipulation or "deceit through humanization" (Butnaru, 2018; cf. Zawieska, 2015). Critical voices caution that, with increasing complexity of the technology, the "connection between input (the programmer's command) and output (how the robot behaves) will become increasingly opaque to people, and may eventually be misinterpreted as free will" (LaFrance, 2016), and that this might cause people to believe that "somebody is at home" in a robot (Scheutz, 2012, p. 3). Some interpret this as the deliberate induction of a false mental model of the robot in the user, exploiting the fact that, for non-experts, a robot can be a black box – or even "indistinguishable from magic"[1] (Clarke, 1973, p. 38).

1 The complete quote by science fiction author Arthur C. Clarke reads: "Any sufficiently advanced technology is indistinguishable from magic" (1973, p. 38).

This criticism is backed by current examples, such as a chat bot being misunderstood as being alive and sentient (Mitsuku Chatbot, 2019), or a computer generated video of a "robot uprising" being mistaken as real (Koebler, 2019).

Even when users are aware that a robot is in fact not alive, and merely attribute animacy to it on a playful, metaphorical level, this can have palpable emotional consequences. There are many examples of people grieving about "dying" robots – such as the Philae asteroid lander (Feltman, 2014; cf. Chapter 4), EOD robots destroyed on duty (e.g. J. Carpenter, 2016; Garreau, 2007), the service robot Jibo (Carman, 2019), or the robot dog Aibo (Griffin, 2015). On a practical level, attributing animacy to robots might make humans hesitant to "abuse" or "kill" robots (Bartneck et al., 2007; A. M. Rosenthal-von der Pütten et al., 2013; Sandry, 2018; James Vincent, 2019), or to deploy them into dangerous situations (J. Carpenter, 2016, p. 44; Sandry, 2015b, p. 106). The "unidirectional emotional bonds" (Scheutz, 2012) between humans and robots could also be exploited – for example by pressuring users into buying updates for their robot in order to keep it "alive". Some authors view these phenomena as expressions of an overall loss of authenticity in our technologized society, warning of illusory experiences replacing genuine relationships (Sparrow, 2002), or diagnosing a "Culture of Simulation" (Turkle, 1997), populated by machines designed in such a way that they make us "fool ourselves" (Turkle, 2011a, p. 20).

Moreover, a mental model of robots as animate – even just on a metaphorical level – sets high expectations for their physical and interactive abilities. Potential robot users have been found to have a strong expectation bias towards how robots are represented in the media. This includes the expectations that "representative robots" (T. Nomura et al., 2005, p. 125) have humanlike cognitive abilities (Kriz et al., 2010), are capable of fluent cooperative behavior (Oestreicher & Eklundh, 2006), and have a humanoid physical appearance (T. Nomura et al., 2005). It appears there is a "mismatch" or "conflict between the expectations of the users (that are primarily shaped by movies and fiction), the goals of HRI research, and the needs of the users" (Sandoval, Mubin, & Obaid, 2014, p. 61). When confronted with the actual current state of robot technology – which cannot yet hold up to these high expectations – disappointment can be the consequence. For example, many customers who bought the humanoid Pepper – advertised as a highly interactive entertainment and customer service robot – were so disappointed by its performance that they "fired" their Pepper robot (e.g. Forrest, 2018). Pepper is not the only example of misleading forms of animacy attribution having tangible economic consequences: Robotics and AI businesses have been accused of fostering a "bullshit-industrial complex" (Mallazzo, 2019), deceiving investors and investors into "[throwing] disproportionate amounts of money [at] business ideas that are flat-out unfeasible and incorrectly ambitious" (Montani, 2017).

Critics warn that the narratives and practices we explored in the previous chapters – routinely treating robot technology as quasi-animate beings over and over – push certain ideas about the role of robotics in our future: The idea that robots inevitably will play a crucial role at all (cf. Bischof, 2017a, p. 137); the idea that robots will reach a certain sophistication within a certain time span, like the prediction that by 2050 a team of robots will be capable of winning the human soccer World Cup (Robocup.org, n.d.); the idea that robots will be malicious, even destroy human life – which, a study claims, is held by more than two thirds of UK adults (Business Wire, 2017); and, crucially, the idea that those robots will have a humanoid form (The Royal Society, 2018). Altogether, critics warn, these biased expectations might "affect public confidence and perceptions [and] contribute to misinformed debate" (The Royal Society, 2018, p. 4).

Another concern is that attributions of animacy to robots propagate misconceptions about the current state of robot technology by backgrounding the contribution of human actors to the actions of robots, while at the same time inflating robots' ability to act autonomously. This effectively ignores, or even negates, the complexity and social thickness of the construction of technological systems (Jasanoff & Kim, 2015, p. 2; cf. Bijker, Hughes, & Pinch, 1987; R. Williams & Edge, 1996). There are numerous examples: Science communicators staging space probes as autonomously acting "explorers", thereby taking away well-deserved credit from human scientists and engineers behind the mission (cf. Clancey, 2006); reports framing robots as the perpetrator of accidents, shifting away the blame from the human who actually made a mistake in the programming or control of the robots; reports blaming a medical service robot for delivering news of a terminal illness to a patient, shifting away the blame from the doctor who made the decision to convey the news via the remote-controlled telepresence robot (cf. Becker, 2019); or the news embedding increased automation in a narrative of "robots are coming to take away our jobs", thus not only omitting that it is humans who make the decision to replace human workers with technology, but also fostering the idea of an inevitable robotized future, which humans only can await passively and helplessly (cf. Merchant, 2019).

The Systemic Level

A worry that pervades all these points of criticism is that misconceptions about the current state of the art in robotics, as well as biased expectations about the future of robot technology, may not only have an impact on how individuals perceive and interact with robot technology. Policy makers and academic experts, too, fall for the biased representations and science fiction-inspired narratives in science communication and the media. Critical voices thus warn that even far-reaching policy decisions are at risk of being made based on misconceptions about robot technology.

Indeed, practices of animacy attribution to robotics can be found in political contexts as well. Obvious at first glance is a strong propensity for using humanoid robots to visualize not only robotics topics, but also neighboring areas such as artificial intelligence. For example, the EU Parliament does not only use the expression "rise of the robots" on their news website (e.g. European Parliament News, 2016), it also features pictures of fictional androids in marketing materials, like those for a hearing on the legal and ethical aspects of robotics and artificial intelligence (see Figure 10). The U.S. Department of Defense features cute humanoid cartoon robots in the logo for their Algorithmic Warfare Cross-Functional Team (see Figure 10) – whose work does in fact not focus on robots, but on computer vision (Pellerin, 2017; G. L. Scott, 2018). And the German Bundestag's Enquete Commission on Artificial Intelligence uses pictures of humanoid robots to illustrate news articles on all kinds of AI topics on their website (see Figure 10).

A strong humanoid bias can also be observed in the context of political events. Especially the practice of featuring commercial humanoid robots as "guest speakers", staging them as sentient, autonomously acting beings, has drawn considerable criticism. For example, a Pepper robot was "invited" to "speak" in the UK parliament as a "witness expert" for robotics and AI topics (UK Parliament, 2018; see Figure 11). The event was quickly criticized by the robotics community as a publicity stunt and even potentially illegal practice (Bryson, 2018a; Volpicelli, 2018).

Hanson Robotics' Sophia robot is probably the most famous – or rather, infamous – robot in the political arena. The robot has been staged as a "speaker" at various political events – among them the 2018 Munich Security Conference (see Figure 11) and several United Nations conferences (ECOSOC, 2017; UNDP, 2018). Sophia has also been at the center of several marketing stunts in the political context: She was named the United Nations Development Programme's first-ever non-human Innovation Champion (UNDP, 2017), was made a honorary citizen of Saudi Arabia (Sini, 2017), and was issued an Azerbaijani visa (Armstrong, 2018).

Figure 10: Top left – Poster for a Hearing on Legal and Ethical Aspects of Robotics and Artificial Intelligence in the European Parliament (2016). Top right – News article on the logo of the US Department of Defense's Algorithmic Warfare Cross-Functional Team (2018). Bottom – News posts illustrated with humanoid robots on the website of the German Bundestag's Enquete-Commission on Aritificial Intelligence (2019).

Sources: http://www.europarl.europa.eu/committees/en/juri/events-hearings.html?id=20160421CHE00181 (top left, accessed on 2019-10-12) | https://www.inverse.com/article/45423-project-maven-logo-department-of-defense-google (top right, screenshot taken on 2019-12-07) | https://www.bundestag.de/ausschuesse/weitere_gremien/enquete_ki (bottom, screenshots taken on 2019-10-12).

At all of these, and the many other events where robots "spoke", it was never clearly disclosed who authored the robots' statements and to which extent the companies providing the robots were involved (Cuthbertson, 2018). Instead, the robots were presented as if they were animate and speaking for themselves.

Figure 11: Left – Newspaper article about a Pepper Robot in the UK Parliament (2018). Right – Robot "Sophia" at the 2018 Munich Security Conference.

Sources: https://www.dawn.com/news/1439541 (left, screenshot taken on December 7, 2019). | https://commons.wikimedia.org/wiki/File:Wolfgang_Ischinger_mit_Roboter_Sophia_MSC_2018.jpg (right, accessed 2019-10-12). Author: MSC/Kuhlmann. Image available under the CC BY 3.0 DE license (https://creativecommons.org/licenses/by/3.0/de/deed.en).

The way these commercial humanoid robots are "paraded around" at political events draws consistent criticism, especially from the robot and AI ethics community. Critics fear that marketing stunts like Sophia's performances, by presenting a biased and distorted image of the current state of technology, make it difficult for government and policy actors to ground their decision on sound facts (e.g. Sharkey, 2018). Fernaeus and colleagues (2009) even warned of a "robot cargo cult" (cf. Feynman, 1974), in which unproven ideas are presented as facts. Crucially, misconceptions about the current state of the art caused by this kind of robot "marketing" are not limited to non-experts. Also funding decisions for robotics research and development, made by reviewers who should be aware of the current state of technology, are sometimes heavily influenced by the ubiquity of narratives of animate (appearing) robots in public discourse. Robotics professor Tony Belpaeme (2018) reported that "an EU project reviewer express[ed] disappointment in [Belpame's team's] slow research progress, as the Sophia bot clearly showed that the technical challenges [they] were still struggling with were solved" – the reviewer apparently having fallen for Hanson Robotics' well-staged demonstrations.

Attributions of animacy to robots are not only discussed critically for their potential to cause misconceptions about current technology. Another major area of concern is that systematically biased representation of robots as animate "may sustain and trigger unrealistic visions", that "not only the general public, but also researchers may maintain an unrealistic, even fantasy-based, perspective of what robots are and could be" (Fernaeus et al., 2009, p. 279), and that not making it clear that even a human-like robot's behavior is controlled by humans "might lead us to design legislation based on the form of a robot, and not the function, … a grave mistake" (Richards & Smart, 2013, p. 21).

In the context of political discourse, including the discourse surrounding funding initiatives for robot technology development, as well as robotics legislation, references to robot animacy are not only observable in the ubiquity of humanoid robot illustrations and marketing stunts. "Science fiction and fantasy are increasingly invoked by policy elites in service of arguments about the real world" (C. Carpenter, 2016, p. 53), serving as either guiding visions or deterrent scenarios. As discussed in Chapter 1, research agendas and roadmaps for robotics innovation, for example by the European Commission, often draw motivation and justification from science fiction-inspired visions of the future, featuring scenarios of interactive, even human-inspired, robot companions and coworkers.

However, not only utopian science fiction narratives can be observed in political discourse. Dystopian scenarios, such as the Terminator movies, are frequently exploited in the controversial discussion of armed conflicts and the development of autonomous weapons (Sharkey, 2018). These narratives are kindled by prominent activists of the anti-autonomous weapons movement, such as Stephen Hawking or Elon Musk, "to signal other broadly recognized meanings, such as the perceived potential impending crisis of an enormous magnitude if these systems are widely used" (J. Carpenter, 2016, p. 24; cf. Gibbs, 2014; Mick, 2014). Critical voices caution that this could "mislead the public on the actual dangers of artificial intelligence" (Shead, 2019). References to science fiction also fall on fertile ground in the military community: A representative of the US Pentagon explicitly stated he is concerned about robots becoming like "a Terminator without a conscience" (Silver, 2016).

Criticism of a misguided use of science fiction narratives in the context of legislation is especially present in the discourse around the legal status of robots. In discussions on who should carry the legal responsibility for accidents caused by robots, Isaac Asimov's (1950) Three Laws of Robotics[2] are almost routinely used as a base for discussion, or even as an explicit model for robot legislation (Murphy & Woods, 2009). The European Parliament's "Resolution on Civil Law Rules of Robotics" makes explicit reference to Asimov's Three Laws having to be upheld (2017, p. 4). Critics caution against using a literary plot device as a basis for legislative decisions. The Three Laws are, after all, formulated deliberately vague so they can be broken to drive the story forward (P. W. Singer, 2009, p. 520). Additionally, Asimov's Laws are – at least as of yet – technologically impossible to "install" in a robot (ibid.). For the robotics community itself, they are "little more than an imaginative literary device" (McCauley, 2007, p. 159). Even a study commissioned by the European Parliament itself criticized the explicit references to the Three Laws in a EU policy document (2016, p. 12) and noted that "when we consider civil liability in robotics, we come up against fanciful visions about robots. Here we must resist calls to establish a legal personality based on science fiction" (ibid., p. 5). The UNESCO World Commission on the Ethics of Scientific Knowledge and Technology warns against "call[ing obots] 'persons' as long as they do not possess some additional qualities typically associated with human persons, such as freedom of will, intentionality, self-consciousness, moral agency or a sense of personal identity" (COMEST, 2017, p. 46). An Open Letter signed by "Artificial Intelligence and Robotics experts, industry leaders, law, medical and ethics experts" criticizes the "bias based on an overvaluation of the actual capabilities of even the most advanced robots ... and a robot perception distorted by Science-Fiction and a few recent sensational press announcements" (Robotics-Openletter.eu, n.d.).

The focus of public, political, and legislative discourse on a narrative of robots as futuristic, animate-appearing, and humanlike – be it inspired by

2 The Three Laws of Robotics: "(1) A robot may not injure a human being or, through inaction, allow a human being to come to harm. (2) A robot must obey the orders given it by human beings except where such orders would conflict with the First Law (3) A robot must protect its own existence as long as such protection does not conflict with the First or Second Laws and (0) A robot may not harm humanity, or, by inaction, allow humanity to come to harm" (Asimov, 1950).

science fiction or by other representations of robots – has another problematic consequence. Critics warn that it shifts away attention from other equally, if not more, important aspects of robotics and automation. In this context, Chihyung Jeon (2016) described robot technology as a "technofuturistic escape": By promoting idealized scenarios of a future in which today's pressing problems have been solved by robot technology, policy-makers are able to evade having to address current problems. The ubiquitous references to a dystopian robot future are criticized for deflecting attention from the fact that robots, but also other non-embodied autonomous technologies, are already inherently embedded in, and have impact on, our current lives:

> "For all the fears of world where robots rule with an iron fist, we already live in a world where machines rule humanity in another way. … We're embedded in a matrix of technology that increasingly shapes how we live, work, communicate, and now fight." (P. W. Singer, 2009, p. 515)

A 2018 cartoon commented on the apparent lack of interest in short- and medium term consequences of artificial intelligence and robotics by creating an imaginary timeline of the infamous AI apocalypse (see Figure 12).

Figure 12: Cartoon "Robot Future" (XKCD, 2018).

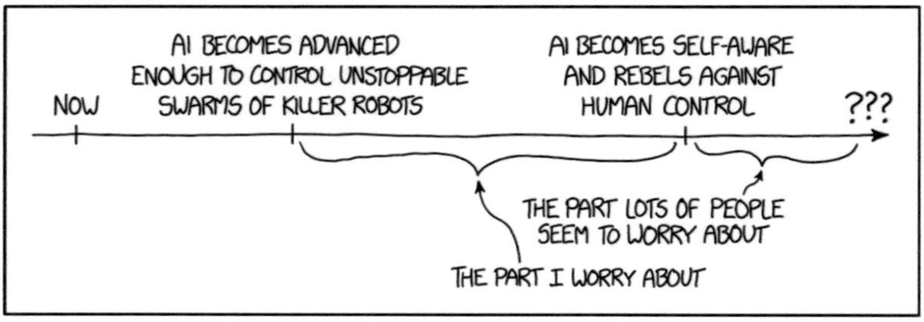

Source: https://xkcd.com/1968 (accessed on 2019-10-13). Image used in accordance with the artist's guidelines (https://xkcd.com/license.html).

The prominent narrative of "robots are coming to take away our jobs" is criticized not only for omitting that it is humans who make the decision to replace human workers with technology, but also for fostering the idea of an inevitable robotized future that humans only can await passively and helplessly (cf. Merchant, 2019). Moreover, the constant discourse on robots as autonomous and agentic is feared to leave the impression that it is they who are responsible for developments such as technological unemployment: "It is easier and more compelling to imagine humanoid robots than to consider the evolution of the consequences for business models, organizations and labour" (Craig, 2019, p. 40). In reality, of course, it is humans who make the decision to automate traditionally human tasks. In other words: "'Robots' Are Not 'Coming for Your Job'—Management Is" (Merchant, 2019). Neither are accidents caused by robots the fault of the robot individual. After all, "robots are simply tools of various kinds, albeit very special tools, and the responsibility of making sure they behave well must always lie with human beings" (Boden et al., 2017, p. 125).

Of course, it is easier to focus on tangible technologies, such as humanoid robots, than on more abstract concepts, such as "algorithms", "big data", or "machine learning". However, it is these complex, non-embodied technologies, which already play an important role in our lives: "There has been too much talk about interesting but irrelevant future questions, and not enough about harder current ones" (Mulgan, cited in Highfield, 2019). This sentiment is shared by many commenters from the AI and robot ethics community:

> "The 'robot invasion' is not something that will transpire as we have imagined it in our science fiction, with a marauding army of evil-minded androids either descending from the heavens or rising up in revolt against their human masters. It is an already occurring event with machines of various configurations and capabilities coming to take up position in our world through a slow but steady incursion." (Gunkel, 2020, p. 1)

6.4. In/Animacy: Beyond Robotics

In the larger context of our current technologized society, among the many perspectives one can choose to explore how technology influences our private and professional lives, robots are an especially tangible and engaging, often even spectacular example. With their long cultural history and their shining roles in fictional narratives, robots are a constant presence. This is only heightened by the current "robotics hype", which places robots at the center of not only significant economic developments but also academic and public discourses. The underlying issues, however, which drive our fascination with robots and feed the ongoing discourses, are not necessarily unique to the topic of robot technology. Some of the more specific aspects observed in this book, such as the constructive function of in/animacy attributions, can be found mirrored in other contexts of our technologized society as well.

As discussed in Chapters 1 and 5, not only is there no professional consensus on what actually "counts" as a robot, also public discourse tends to group robots with what is perceived as neighboring technologies. This includes whole fields, such as artificial intelligence or machine learning, but also specific pieces of technology, such as autonomous cars, drones, or smart home appliances. In the context of these technological fields and artifacts we can observe a similar juggling of ontological categories and, moreover, a thematic overlap with the public and critical discourse on robot technology.

Many of the observations we made specifically for robotics can be transferred to the context of artificial intelligence – including much of the critical discourse (Kurenkov, 2019; Marcus, 2013; Schwartz, 2018; Togelius, 2017). Users of virtual assistants like Alexa[3] or Siri[4] can develop emotional connections to the software personas (Aronson & Duportail, 2018), going so far that they can be more open and willing to disclose personal feelings to virtual humans, compared to real humans (Lucas et al., 2014). Outside the context of deliberately socially interactive AI, we can observe further practices of personification. Typical phrases with umbrella terms insinuating cognitive processes, such as "teaching an AI to do something" or "the AI thinks that", reveal practices of animacy attribution. The backgrounding of human involvement, too, is an issue in the context of artificial intelligence, for example in the case of "pseudo-AI". There are reports of companies charging customers for the services of "AI assistants" – which are in fact nothing but human workers pretending to be the AI by communicating in a "robotic" style (e.g. Shane, 2018; Solon, 2018). Newspapers were accused of "faking" after they published an article presumably written by a neural net, while in fact human journalists were involved in the editing process (e.g. Seabrook, 2019; Lowndes, 2020). We can also observe cases of agency and intentionality being attributed to algorithms, in order to shift away the blame for questionable practices from human developers and management – such as the discussion about a credit algorithm "deciding" that female customers were less credit-worthy (Heinemeier Hansson, 2019; Vigdor, 2019).

3 https://developer.amazon.com/en-US/alexa (accessed 2020-01-08).

4 https://www.apple.com/siri (accessed 2020-01-08).

Outside the realm of autonomous technologies we find further examples of similarly constructive attributive practices scraping, even crossing the borders of ontological categories like "human" and "nonhuman", "artificial" and natural".

Lucy Suchman (2011), for example, discussed the concept of the model organism – meaning any nonhuman species serving as a biological research platform, with the crucial expectation that scientific insights made in the model organism can in some way be transferred from this organism to another one (cf. Fields & Johnston, 2005). Typical model organisms are the *E. coli* bacterium, laboratory rats and mice, or the common fruit fly *Drosophila*. In the work of researchers like Robert Kohler (1994) or Lynda Birke and colleagues (2004) we can find several parallels to the case of robot technology. In their unique role as models for other organisms, these species are an example for human characteristics being mapped onto nonhuman entities, in that – at least in public discourse – the crucial differences between the animal and human organism is played down. Preclinical research on mice and rats is frequently "hyped" when making its way into public discourse, reports misleadingly making the results appear directly transferable to human organisms (Heathers, 2019). Model organisms are also an example for living entities being reconstructed, at the same time, as research tools and as active participants in the knowledge making process. In a similar vein, Karin Knorr-Cetina (1997, 1998) described how cytogeneticist Barbara McClintock perceived herself to be "among the chromosomes" during her work, and how "she not only identif[ied] with them, she enter[ed] their environment, in which she bec[ame] situated as 'one of them'" (Knorr-Cetina, 1997, p. 24). Here, similarly to what we observed for the case of robots, the practice of identifying with the chromosomes, indirectly giving the objects of research the status of a persona, was constructive for McClintock's work process.

On a more general and abstract level, we can observe more parallels to the practices and discourses discussed in this book. The issue of ontological categorizations at the border of the "technical" and the "natural" is a central point of discussion in the context of how we perceive and represent robot technology. In the case of robotics, the central question is whether an objectively technical, inanimate object shares enough characteristics with "traditionally animate" entities to be sorted in the same ontological category, or merits the creation of a completely new category, which would effectively break the traditional dichotomy of "animate" and "inanimate".

In other contexts, we can find similar ontological questions. One example is the case of so-called biofacts – artificially created biotic artifacts, such as cloned animals, artificially grown body tissues, or genetically modified fruit. Biofacts are (or were at a certain point) phenomenologically animate[5], but their development and growth processes are technologically controlled (Gill, Torma, & Zachmann, 2018). In the case of robot technology, the artificial-technological aspects of the artifact are relatively obvious on a phenomenological level. With the exception of extremely realistic androids, robots are usually identifiable technological artifacts at first glance. In the case of biofacts, on the other hand, their artificiality and the technological influence that shaped them are invisible – sometimes even down to the molecular level (Karafyllis, 2003). Both robot technology and biofacts, however, are situated at or on the border of the natural and the technological, making their assignment to traditional ontological categories difficult or even impossible. As discussed in the introductory chapter (Section 1.1), robots are sometimes assigned to a completely new category, somewhere between "animate" and "inanimate". Likewise, biofacts are discussed to be a new category for themselves, ontologically located between "artifact" and "animate being" (Karafyllis, 2003, p. 16).

Robots are not the only technology that, in the words of Donna Haraway, "has made thoroughly ambiguous the difference between natural and artificial" (1991, p. 152). However, robots are a highly instructive case to observe how our technologized society constantly forces us to question seemingly long-established ontological boundaries. In this sense, the constant switching of attributions we observed across all contexts along the life cycle of robots also is a manifestation of the negotiation of these ambiguous boundaries between us and technology, and a negotiation of how much control and closeness we are willing to allow technology – both on a practical and on an emotional level.

6.5. Speaking Clearly: A Take-Home Message

Over the course of this book, on our long and winding journey along the life cycle of robots, we encountered different practices of talking and thinking about robots – these peculiar machine-beings that seemingly only recently

5 Here, the notion of animacy is associated with the existence of biological growth processes.

stepped from factories and fiction into in our physical lives. We found that all kinds of people – from experts who work closely with state-of-the-art robot technology, to lay people who only know robots from science fiction movies – routinely and effortlessly balance and play with robots' complex and confusing ontological status. It appears that humans are able to see robots as both, inanimate machines and animate beings, and able to express both perspectives in the way they talk about and interact with robots, without it feeling contradictory. We saw that the practice of balancing and playing with robots' in/animacy has crucial, constructive, and useful functions. However, we also found that this practice – if practiced unthinkingly or too opportunistically – is perceived as causing problematic consequences.

At the end of this book, let me take a step back from the position of the scientific observer and consider how some insights of my research can be applied constructively to our current and future lives in a robotized society.

The first insight is that attributions of animacy to robots – and to other autonomous (appearing) technologies – are a ubiquitous, persistent, and very old phenomenon. It is deeply ingrained in our cognitive-perceptual system and, in all likelihood, we will not be able to "stop it". In other words: "Just telling people not anthropomorphize robots won't work" (De Graaf, 2017). In fact, we do not have to stop: As we saw, we are doing an excellent job at juggling, on the one hand, our rational knowledge about the inanimacy of a robot and, on the other hand, the playful metaphorical attributions that help us interact with robots and communicate with other humans about robots in a meaningful way. For us, "these seemingly contradictory features – a thing and a living creature – [can] unproblematically coexist" (Alač, 2016, p. 12). Our ability to switch effortlessly between different ontological perspectives on technology serves us well in the technologized society we live in. In this sense, attributions of in/animacy are a cognitive, practical, and discursive tool that helps us make sense of complex autonomous technologies and the different contexts where we encounter them. Maybe we can "cut each other, and ourselves, some intellectual slack when it comes to [these] familiar, relatively benign, kinds of self-indulgence ... [they] can co-exist with ordinary honesty and commitment to truth" (Blackford, 2012, p. 50).

This commitment to truth is crucial, however. Knowing now how powerful even "only" metaphorical attributions of animacy can be, which very concrete consequences they can have – such as when legislative decisions are based on them – we should take them seriously. For those of us who talk about, who write about, who present robot technology, this does not mean that we should not use those playful metaphors and references to science fiction. As we learned, they are inevitable in a way, they are useful – and they are fun. That is, they are as long they are applied mindfully and responsibly.

We, the readers and the writer of this book, are now experts for the issue of in/animacy attributions to robots. But not everyone is. Not everyone is able to assess whether an extremely realistically behaving humanoid robot is remote controlled or whether it has "real" intentionality, intelligence, and animacy. Not everyone knows that "killer robots" do not look like the Terminator. It is for these people we need to make an effort not to let opportunism turn playful attributions of animacy into deception. What does this mean in practice?

On a fundamental level, it means being aware that technology and society are always entangled. It means knowing that robotics, like any other technology, can never be "neutral" and unbiased, as its production is always inherently connected to its societal context (e.g. Jasanoff, 2004).

On a practical level, it means accurately describing the current state of robot technology; clarifying what is fact and what is fiction, and separating the present from an imaginary future; refraining from making non-experts believe that a robot is more intelligent and autonomous than it actually is; being upfront about the influence of humans on robot's behavior, and about the limitations of a robot's capabilities (Kurenkov, 2019). It also means using realistic pictures as illustrations for articles – or at the very least, providing explanatory image captions; clarifying that the technology described in an article does not look like the humanoid robot in the picture, but that the picture is from a science fiction movie.

More generally, it also means not letting the "exoticness" of robot technology distract us from other equally, or even more, pressing issues of technological innovation. So, instead of only wondering whether Boston Dynamics' robots will bring about the robot apocalypse, let us also talk about them being funded by the Department of Defense, about the use of their robots for surveillance and in law enforcement (Cuthbertson, 2019; Schwartz, 2018; Sullivan, Jackman, & Fung, 2016). While worrying about the dangers of Terminator-like autonomous weapons, let us also consider the dangers of embedded forms of intelligence in "smart homes" and "smart cities" (Craig, 2019, p. 40; König, 2019). And when following the exciting adventures of cute space robots, let us also consider the ethical, environmental, and economic consequences of space exploration and planetary exploitation (L. Wright, 2016).

Overall, it means taking a pragmatic approach to existing and future technologies – robotic and otherwise (cf. von Gehlen, 2018). To promote one's own and others' technology literacy[6] by being open for emerging technologies, but also to keep in mind their limitations (cf. Renn, 2011).

And finally, it means to not let science fiction make us think that one or the other robotic future is inevitable, that the machines are in control, but to be aware of our power to influence the presence of robot technology in our present and future lives.

> "The machine is not an it to be animated, worshipped, and dominated. The machine is us, our process, an aspect of our embodiment. We can be responsible for machines; they do not dominate or threaten us. We are responsible for boundaries; we are they." (Donna Haraway, "A Cyborg Manifesto", 1991, p. 81)[7]

6 A loose translation of the German term "Technikmündigkeit", "Mündigkeit" meaning maturity or majority. Hat tip to Ilja Sperling for suggesting this translation.

7 Hat tip to Beth Singler (2019) for using this quote in a blog post.

References

Aarts, H., Dijksterhuis, A., & Dik, G. (2013). Goal Contagion. In J. Y. Shah & W. L. Gardner (Eds.), *Handbook of Motivation Science* (pp. 265–280). New York; London: Guilford Publications.

Acatech. (2012). *Perspektiven der Biotechnologie-Kommunikation: Kontroversen – Randbedingungen – Formate.* Wiesbaden: Springer Vieweg.

Akrich, M. (1995). User Representations: Practices, Methods and Sociology. In A. Rip, T. J. Misa, & J. Schot (Eds.), *Managing Technology in Society. The Approach of Constructive Technology Assessment* (pp. 167–184). London; New York: Pinter Publisher.

Alač, M. (2016). Social Robots: Things or Agents? *AI & Society*, *31*(4), 519–535. https://doi.org/10/ggbn6t

Alač, M., Movellan, J., & Tanaka, F. (2011). When a Robot is Social: Spatial Arrangements and Multimodal Semiotic Engagement in the Practice of Social Robotics. *Social Studies of Science*, *41*(6), 893–926. https://doi.org/10/c8jkpn

Alba, D. (2016). Root is a Little Robot on a Mission to Teach Kids to Code. *WIRED.* https://www.wired.com/2016/04/harvard-built-robot-teach-kids-code (accessed 2017-02-21).

Alpeyev, P., & Amano, T. (2016). Softbank's Struggles with Pepper Keep Son's Robot Dreams on Hold. *The Economic Times.* http://economictimes.indiatimes.com/small-biz/security-tech/technology/softbanks-struggles-with-pepper-keep-sons-robot-dreams-on-hold/articleshow/55108707.cms (accessed 2016-10-31).

Ambasna-Jones, M. (2015). Will Automation and the Internet of Things Lead to Mass Unemployment? *The Guardian.* https://www.theguardian.com/media-network/2015/may/27/internet-of-things-automation-unemployment (accessed 2015-02-22).

Ambasna-Jones, M. (2016). How Social Robots Are Dispelling Myths and Caring for Humans. *The Guardian*. https://www.theguardian.com/media-network/2016/may/09/robots-social-health-care-elderly-children (accessed 2016-02-22).

Angle, C. (2019). When iRobot created Roomba, we didn't want it to be cute (…) [Tweet by @colinangle]. https://twitter.com/colinangle/status/1183723104176095238 (accessed 2016-10-19).

Armstrong, P. (2018). AI Humanoid "Sophia" is Granted First Ever Robot Visa, Speaks with President. *Forbes*. https://www.forbes.com/sites/paularmstrongtech/2018/10/27/ai-humanoid-sophia-is-granted-first-ever-robot-visa-speaks-with-president/#5e4d45461f39 (accessed 2018-10-31).

Aronson, P., & Duportail, J. (2018). Can Emotion-Regulating Tech Translate Across Cultures? *AEON*. https://aeon.co/essays/can-emotion-regulating-tech-translate-across-cultures (accessed 2018-12-09).

Arthur, W. B. (2010). *The Nature of Technology: What it is and How it Evolves*. London: Penguin.

Asimov, I. (1950). Runaround. In *I, Robot* (p. 40 pp.). New York: Gnome Press.

Astro Robonaut. (2016). It was a great pleasure meeting @georgetakei (…) [Tweet by @AstroRobonaut]. https://twitter.com/AstroRobonaut/status/771806489132556292 (accessed 2016-07-17).

Austen, J. (1813). *Pride and Prejudice*. Whitehall: T. Egerton.

Automatones. (n.d.). *The Theoi Project*. https://www.theoi.com/Ther/Automotones.html [sic] (accessed 2019-10-26).

Badham, J. (1986). *Short Circuit* [Film]. TriStar Pictures.

Bankston, K., & Finn, E. (2019, July). *The Sci-Fi Feedback Loop*. Presented at the Conference "What Sci-Fi Futures Can (and Can't) Teach Us About AI Policy." http://opentranscripts.org/transcript/sci-fi-feedback-loop (accessed 2020-09-22).

Barrett, J. L., & Lanman, J. A. (2008). The Science of Religious Beliefs. *Religion*, *38*(2), 109–124. https://doi.org/10/fntfpr

Bartneck, C., Bleeker, T., Bun, J., Fens, P., & Riet, L. (2010). The Influence of Robot Anthropomorphism on the Feelings of Embarrassment when Interacting with Robots. *Paladyn*, *1*(2), 109–115. https://doi.org/10/ftqsx4

Bartneck, C., & Forlizzi, J. (2004). A Design-Centered Framework for Social Human-Robot Interaction. *RO-MAN 2004*, 591–594. https://doi.org/10/fvq5wk

Bartneck, C., Kanda, T., Ishiguro, H., & Hagita, N. (2009). My Robotic Doppelgänger – A Critical Look at the Uncanny Valley. *RO-MAN 2009*, 269–276. https://doi.org/10/cfhp69

Bartneck, C., Kanda, T., Mubin, O., & Al Mahmud, A. (2009). Does the Design of a Robot Influence Its Animacy and Perceived Intelligence? *International Journal of Social Robotics*, *1*(2), 195–204. https://doi.org/10/dj4d63

Bartneck, C., Kulić, D., Croft, E., & Zoghbi, S. (2009). Measurement Instruments for the Anthropomorphism, Animacy, Likeability, Perceived Intelligence, and Perceived Safety of Robots. *International Journal of Social Robotics*, *1*(1), 71–81. https://doi.org/10/bvsbrx

Bartneck, C., Nomura, T., Kanda, T., Suzuki, T., & Kato, K. (2005). A Cross-Cultural Study on Attitudes Towards Robots. *HCI International 2005*. http://www.bartneck.de/publications/2005/nars/index.html (accessed 2020-09-22).

Bartneck, C., Suzuki, T., Kanda, T., & Nomura, T. (2007). The Influence of People's Culture and Prior Experiences with Aibo on Their Attitude Towards Robots. *AI & Society*, *21*(1–2), 217–230. https://doi.org/10/ddbb2t

Bartneck, C., van der Hoek, M., Mubin, O., & Al Mahmud, A. (2007). "Daisy, Daisy, Give Me Your Answer Do!": Switching Off a Robot. *HRI '07*, 217–222. https://doi.org/10/bbvxhd

BBC News. (2019). French Army Asks Sci-Fi Writers to Predict Future. *BBC*. https://www.bbc.com/news/world-europe-49044892 (accessed 2019-07-29).

Becker, R. (2019). Why Hospitals Shouldn't Use Telepresence Robots to Deliver Devastating News. *The Verge*. https://www.theverge.com/2019/3/11/18256929/telepresence-robot-hospitals-patient-information-medical-news-ethics-kaiser (accessed 2019-03-15).

Beckert, J. (2016). *Imagined Futures: Fictional Expectations and Capitalist Dynamics*. Cambridge, MA: Harvard University Press.

Belfiore, M. (2014). The Robots are Coming. *AEON*. http://aeon.co/magazine/technology/meet-darpas-new-generation-of-humanoid-robots (accessed 2014-04-10).

Belpaeme, T. (2018). I had an EU project reviewer express disappointment in our slow research progress (...) [Tweet by @TonyBelpaeme]. https://twitter.com/TonyBelpaeme/status/1064166510103474176 (accessed 2018-11-21).

Bethel, C. L., & Murphy, R. R. (2006). Affective Expression in Appearance Constrained Robots. *HRI'06*, 327–328.

Bethel, C. L., & Murphy, R. R. (2010). Review of Human Studies Methods in HRI and Recommendations. *International Journal of Social Robotics*, 2(4), 347–359. https://doi.org/10/fm789t

Biggs, J. (2016). Our Robotic Overlords Have Written Us a Christmas Carol! *TechCrunch*. http://social.techcrunch.com/2016/12/09/our-robotic-overlords-have-written-us-a-christmas-carol (accessed 2017-04-10).

Bijker, W. E., Hughes, T. P., & Pinch, T. J. (Eds.). (1987). *The Social Construction of Technological Systems: New Directions in the Sociology and History of Technology*. Cambridge, MA: MIT Press.

Bijker, W. E., & Law, J. (Eds.). (1992). *Shaping Technology/Building Society: Studies in Sociotechnical Change*. Cambridge, MA: MIT Press.

Birke, L., Bryld, M., & Lykke, N. (2004). Animal Performances: An Exploration of Intersections between Feminist Science Studies and Studies of Human/Animal Relationships. *Feminist Theory*, 5(2), 167–183. https://doi.org/10/dcgv8f

Bischof, A. (2015). *Wie kommt die Robotik zum Sozialen? Epistemische Praktiken in der Sozialrobotik* (Doctoral dissertation). Technische Universität Chemnitz.

Bischof, A. (2017a). *Soziale Maschinen bauen: Epistemische Praktiken der Sozialrobotik*. Bielefeld: Transcript.

Bischof, A. (2017b). Wie Robotik Zukunft herstellt. *Soziologieblog*. https://soziologieblog.hypotheses.org/10518 (accessed 2017-05-02).

Bischof, A. (2019). *Menschenähnliche Roboter als verhängnisvolle Verheißung – Wissenschaftssoziologische Analyse der Mensch-Roboter-Interaktion*. Presented at the Symposium Junge Akademie, Mainz.

Bischof, A., & Both, G. (2015). *Counting Clicks instead of Citations. Youtube Videos as Scientific Currency*. Presented at the 14th Annual STS Conference, Graz.

Bishop, B. (2014). Hey, Remember That Time Google Accidentally Made Skynet? *The Verge*. http://www.theverge.com/2014/6/24/5835708/google-x-labs-artificial-intelligence-elon-musk-terminator (accessed 2017-04-11).

Black, M. (1979). More About Metaphor. In A. Ortony (Ed.), *Metaphor and Thought* (pp. 19–43). Cambridge, UK; New York: Cambridge University Press.

Blackford, R. (2012). Robots and Reality: A Reply to Robert Sparrow. *Ethics and Information Technology*, 14(1), 41–51. https://doi.org/10/fskbxd

Blomkamp, N. (2015). *Chappie* [Film]. Sony Pictures.

Boden, M., Bryson, J., Caldwell, D., Dautenhahn, K., Edwards, L., Kember, S., … Winfield, A. (2017). Principles of Robotics: Regulating Robots in the Real World. *Connection Science*, 29(2), 124–129. https://doi.org/10/gf9whn

Böhle, K., & Bopp, K. (2013). What a Vision: The Artificial Companion. A Piece of Vision Assessment Including an Expert Survey. *Science, Technology & Innovation Studies*, 10(1), 155–186.

Boland, H. (2019). The Electronic Skin That Could Allow Robots to "Feel" Pain. *The Telegraph.* https://www.telegraph.co.uk/technology/2019/09/02/electronic-skin-could-allow-robots-feel-pain (accessed 2019-09-04).

Borenstein, J., & Arkin, R. (2019). Robots, Ethics, and Intimacy: The Need for Scientific Research. In D. Berkich & M. V. d'Alfonso (Eds.), *On the Cognitive, Ethical, and Scientific Dimensions of Artificial Intelligence: Themes from IACAP 2016* (pp. 299–309). https://doi.org/10/dhtb

Borup, M., Brown, N., Konrad, K., & Lente, H. V. (2006). The Sociology of Expectations in Science and Technology. *Technology Analysis & Strategic Management*, 18(3–4), 285–298. https://doi.org/10/cwqzcz

Bostedt, S. L. (2016). Robots are Taking Over the Food We Eat. No Joke. *The Chicago Tribune.* http://www.chicagotribune.com/redeye/redeye-robots-are-taking-over-the-food-we-eat-no-joke-20160607-story.html (accessed 2017-02-21).

Boston Dynamics. (2016). *Atlas, The Next Generation* [Youtube Video]. https://www.youtube.com/watch?v=rVlhMGQgDkY (accessed 2020-09-22).

Boston Dynamics. (2018a). *Hey Buddy, Can You Give Me a Hand?* [Youtube Video]. https://www.youtube.com/watch?v=fUyU3lKzoio (accessed 2020-09-22).

Boston Dynamics. (2018b). *Parkour Atlas* [Youtube Video]. https://www.youtube.com/watch?v=LikxFZZO2sk (accessed 2020-09-22).

Boston Dynamics. (n.d.). Boston Dynamics Youtube Channel. https://www.youtube.com/user/BostonDynamics (accessed 2020-09-22).

Both, G. (2015). Youtubization of Research. Enacting the High-Tech Cowboy in Video Demonstrations. In E. Stengler (Ed.), *Studying Science Communication* (pp. 24–27). http://eprints.uwe.ac.uk/26680 (accessed 2020-09-22).

Boyd, R. (2019). We got a roomba and i get it now (...) [Tweet by @ryandroyd]. https://twitter.com/ryandroyd/status/1103782256638812161 (accessed 2019-03-08).

Braun-Thürmann, H. (2002). *Künstliche Interaktion: Wie Technik zur Teilnehmerin sozialer Wirklichkeit wird.* Wiesbaden: Westdeutscher Verlag.

Brenton, H., Gillies, M., Ballin, D., & Chatting, D. (2005). The Uncanny Valley: Does It Exist? *HCI 2005.*

Brett, G. (1954). The Automata in the Byzantine "Throne of Solomon." *Speculum*, 29(3), 477–487. https://doi.org/10/ccdwcd

Brooks, R. (2017a). *Domo Arigato Mr. Roboto.* http://rodneybrooks.com/forai-domo-arigato-mr-roboto (accessed 2017-09-19).

Brooks, R. (2017b). *The Seven Deadly Sins of Predicting the Future of AI.* http://rodneybrooks.com/the-seven-deadly-sins-of-predicting-the-future-of-ai (accessed 2017-09-19).

Brooks, R. (2018). This is completely bogus and a total sham (…) [Tweet by @rodneyabrooks]. https://twitter.com/rodneyabrooks/status/1056247845437636608 (accessed 2018-11-20).

Brown, N., Rappert, B., & Webster, A. (Eds.). (2000). *Contested Futures: A Sociology of Prospective Techno-Science*. Aldershot, UK; Burlington, VT: Ashgate.

Bruckenberger, U., Weiss, A., Mirnig, N., Strasser, E., Stadler, S., & Tscheligi, M. (2013). The Good, The Bad, The Weird: Audience Evaluation of a "Real" Robot in Relation to Science Fiction and Mass Media. In G. Herrmann, M. J. Pearson, A. Lenz, P. Bremner, A. Spiers, & U. Leonards (Eds.), *Social Robotics* (pp. 301–310). https://doi.org/10/dfrw

Bryson, J. (2018a). Let's be VERY clear about this. A person or corporation is committing perjury (…) [Tweet by @j2bryson]. https://twitter.com/j2bryson/status/1050332579197571073 (accessed 2019-10-12).

Bryson, J. (2018b). *Putting Sophia on the Bill, Ethically (an Open Letter).* https://joanna-bryson.blogspot.com/2018/11/putting-sophia-on-bill-ethically-open.html (accessed 2018-11-21).

Bucchi, M., & Mazzolini, R. G. (2003). Big Science, Little News: Science Coverage in the Italian Daily Press, 1946-1997. *Public Understanding of Science, 12*(1), 7–24. https://doi.org/10/b54g5x

Buchanan, D. W. (2015). The Robots Aren't Going to Kill You. *The Chicago Tribune.* http://www.chicagotribune.com/news/opinion/commentary/chi-ibm-watson-robots-artificial-intelligence-20150209-story.html (accessed 2017-02-21).

Burton-Hill, C. (2016). The Superhero of Artificial Intelligence: Can This Genius Keep It in Check? *The Guardian.* https://www.theguardian.com/technology/2016/feb/16/demis-hassabis-artificial-intelligence-deepmind-alphago (accessed 2017-02-22).

Business Wire. (2017). Robots: Better Saved for Sci-Fi Believe UK Consumers (SQS Study). *Business Wire.* http://www.businesswire.com/news/home/20170704005001/en/Robots-Saved-Sci-Fi-UK-Consumers-SQS-Study (accessed 2017-07-20).

Butnaru, C. (2018). The problem of appropriate disclosure in robotics IS a BIG issue (...) [Tweet by @katchja]. https://twitter.com/katchja/status/956966313091624961 (accessed 2018-03-20).

Callon, M. (1986). The Sociology of an Actor-Network: The Case of the Electric Vehicle. In M. Callon, J. Law, & A. Rip (Eds.), *Mapping the Dynamics of Science and Technology: Sociology of Science in the Real World* (pp. 19–34). https://doi.org/10/dfr4

Calvert, J. (2006). What's Special About Basic Research? *Science, Technology, & Human Values*, *31*(2), 199–220. https://doi.org/10/cgzvjv

Cameron, J. (1984). *The Terminator* [Film]. Orion Pictures.

Čapek, K. (1920). *Rossumovi Univerzální Roboti (Rossum's Universal Robots)*. https://archive.org/stream/rurrossumsuniver00apekuoft#mode/2up (accessed 2020-09-22).

Caporael, L. R. (1986). Anthropomorphism and Mechanomorphism: Two Faces of the Human Machine. *Computers in Human Behavior*, *2*(3), 215–234. https://doi.org/10/bkxmf2

Caporael, L. R., & Heyes, C. M. (1997). Why Anthropomorphize? Folk Psychology and Other Stories. In R. Mitchell, N. S. Thompson, & H. L. Miles (Eds.), *Anthropomorphism, Anecdotes, and Animals* (pp. 59–73). Albany, NY: Suny Press.

Carman, A. (2019). They Welcomed a Robot into Their Family, Now They're Mourning Its Death. *The Verge*. https://www.theverge.com/2019/6/19/18682780/jibo-death-server-update-social-robot-mourning (accessed 2019-06-25).

Carpenter, C. (2016). Rethinking the Political / -Science- / Fiction Nexus: Global Policy Making and the Campaign to Stop Killer Robots. *Perspectives on Politics*, *14*(01), 53–69. https://doi.org/10/gfvxw2

Carpenter, J. (2013). *The Quiet Professional: An Investigation of U.S. Military Explosive Ordnance Disposal Personnel Interactions with Everyday Field Robots* (Doctoral dissertation). University of Washington.

Carpenter, J. (2016). *Culture and Human-Robot Interaction in Militarized Spaces: A War Story.* Burlington, VT: Ashgate Pub Co.

Carpinella, C. M., Wyman, A. B., Perez, M. A., & Stroessner, S. J. (2017). The Robotic Social Attributes Scale (RoSAS): Development and Validation. *HRI'17*, 254–262. https://doi.org/10/ggbn6q

Cave, S., Coughlan, K., & Dihal, K. (2019). *"Scary Robots": Examining Public Responses to AI*. Presented at the AI, Ethics, and Society Conference, Honolulu, Hawaii. https://doi.org/10/ggbn5s

Cave, S., & Dihal, K. (2019). Hopes and Fears for Intelligent Machines in Fiction and Reality. *Nature Machine Intelligence*, *1*(2), 74. https://doi.org/10/gg bn5k

Chamorro-Premuzic, T. (2016). Why a Robot Could Be the Best Boss You've Ever Had. *The Guardian*. https://www.theguardian.com/media-n etwork/2016/sep/09/robot-boss-best-manager-artificial-intelligence (accessed 2017-02-22).

Chandler, J., & Schwarz, N. (2010). Use Does Not Wear Ragged the Fabric of Friendship: Thinking of Objects as Alive Makes People Less Willing to Replace Them. *Journal of Consumer Psychology*, *20*(2), 138–145. https://doi.o rg/10/bns7fh

Chang, A. (2014). Amazon Robots Speed Customer Orders but May Lead to Fewer Workers. *The Chicago Tribune*. http://www.chicagotribune.com/bus iness/chi-amazon-robots-20141202-story.html (accessed 2017-22-21).

Chatfield, T. (2016). What Does It Mean to Be Human in the Age of Technology? *The Guardian*. https://www.theguardian.com/technology/2016/jan/20 /humans-machines-technology-digital-age (accessed 2017-02-22).

Chin, M. G., Sims, V. K., Clark, B., & Lopez, G. R. (2004). Measuring Individual Differences in Anthropomorphism Toward Machines and Animals. *Proceedings of the HFES Annual Meeting*, *48*(11), 1252–1255. https://doi.org/1 0/fxphwr

Choi, J. J., Kim, Y., & Kwak, S. S. (2014). Are You Embarrassed?: The Impact of Robot Types on Emotional Engagement with a Robot. *HRI'14*, 138–139. ht tps://doi.org/10/gfkv9q

Clancey, W. J. (2006). Clear Speaking about Machines: People are Exploring Mars, Not Robots. *AAAI '06*. https://www.aaai.org/Papers/Workshops/20 06/WS-06-09/WS06-09-012.pdf (accessed 2020-09-22).

Clarke, A. C. (1973). *Profiles of the Future: An Inquiry into the Limits of the Possible*. London: Pan Books.

Coeckelbergh, M. (2018). How to Describe and Evaluate "Deception" Phenomena: Recasting the Metaphysics, Ethics, and Politics of ICTS in Terms of Magic and Performance and Taking a Relational and Narrative Turn. *Ethics and Information Technology*, *20*(2), 71–85. https://doi.org/10/ggbn55

COMEST. (2017). *Report of COMEST on Robot Ethics*. http://unesdoc.unesco.org /images/0025/002539/253952E.pdf (accessed 2020-09-22).

Computing Community Consortium (Ed.). (2016a). *From Internet to Robotics—A Roadmap for US Robotics*. http://cra.org/ccc/wp-content/uploads/sites/2/20 16/11/roadmap3-final-rs-1.pdf (accessed 2020-09-22).

Computing Community Consortium. (2016b). *Next Generation Robotics*. http://cra.org/ccc/wp-content/uploads/sites/2/2016/06/15097-CCC-Next-Gen-Whitepaper-v4.pdf (accessed 2020-09-22).

Coopmans, C. (2011). "Face Value": New Medical Imaging Software in Commercial View. *Social Studies of Science*, *41*(2), 155–176. https://doi.org/10/d3q8vp

Corbyn, Z. (2015). Robots Are Leaving the Factory Floor and Heading for Your Desk – and Your Job. *The Guardian*. https://www.theguardian.com/technology/2015/feb/09/robots-manual-jobs-now-people-skills-take-over-your-job (accessed 2017-02-22).

Craig, C. (2019). *How Does Government Listen to Scientists?* https://doi.org/10/dfsg

Cramer, H., & Büttner, S. (2011). Things that tweet, check-in and are befriended. Two explorations on robotics and social media. *HRI'11*, 125–126.

Curiosity Rover. (2016a). #IAmThankfulFor my 17 cameras, (...) [Tweet by @MarsCuriosity]. https://twitter.com/MarsCuriosity/status/801849397843611648 (accessed 2018-07-09).

Curiosity Rover. (2016b). I'm alone for the holidays, but thanks to kind acts like this (...) [Tweet by @MarsCuriosity]. https://twitter.com/MarsCuriosity/status/809561317224824833 (accessed 2016-12-19).

Curiosity Rover. (2018). Reunited and it feels so good (...) [Tweet by @MarsCuriosity]. https://twitter.com/marscuriosity/status/1003730850695901184 (accessed 2019-01-04).

Cuthbertson, A. (2018). A Robot is Going to Appear Before Parliament to Give Evidence. *The Independent*. https://www.independent.co.uk/life-style/gadgets-and-tech/news/uk-robot-pepper-commons-ai-select-committee-fourth-industrial-revolution-a8575741.html (accessed 2019-11-12).

Cuthbertson, A. (2019). Robot Dogs Being Tested by Police Described as "Terrifying." *The Independent*. https://www.independent.co.uk/life-style/gadgets-and-tech/news/robots-police-dog-spot-boston-dynamics-a9218491.html (accessed 2019-11-27).

Damiano, L., & Dumouchel, P. (2018). Anthropomorphism in Human–Robot Co-Evolution. *Frontiers in Psychology*, *9*(468). https://doi.org/10/gdbbjk

Darling, K. (2012). *Extending Legal Rights to Social Robots*. Presented at the We Robot Conference. http://papers.ssrn.com/abstract=2044797 (accessed 2020-09-22).

Darling, K., Nandy, P., & Breazeal, C. (2015). Empathic Concern and the Effect of Stories in Human-Robot Interaction. *RO-MAN 2015*, 770–775. https://doi.org/10/ggbn6k

Daston, L. (1995). How Nature Became the Other: Anthropomorphism and Anthropocentrism in Early Modern Natural Philosophy. In S. Maasen, E. Mendelsohn, & P. Weingart (Eds.), *Biology as Society, Society as Biology: Metaphors* (pp. 37–56). https://doi.org/10/dfsn

Daston, L. (2000). Historische Überlegungen zum Anthropomorphismus und zur Objektivität in den Wissenschaften. In B. O. Küppers (Ed.), *Die Einheit der Wirklichkeit* (pp. 27–43). München: Fink, Wilhelm.

Dautenhahn, K. (2007). Methodology & Themes of Human-Robot Interaction: A Growing Research Field. *International Journal of Advanced Robotic Systems, 4 (1)*, 103–108. https://doi.org/10.5772/5702

Davies, W. (2016). Robot Amelia – a Glimpse of the Future for Local Government. *The Guardian.* https://www.theguardian.com/public-leaders-network/2016/jul/04/robot-amelia-future-local-government-enfield-council (accessed 2017-02-22).

De Graaf, M. (2016). An Ethical Evaluation of Human–Robot Relationships. *International Journal of Social Robotics, 8*(4), 589–598. https://doi.org/10/ggbn62

De Graaf, M. (2017). Just telling people not anthropomorphize robots won't work (…) [Tweet by @RoboNarratives]. https://twitter.com/RoboNarratives/status/820259413353234438 (accessed 2017-01-16).

De Graaf, M., & Ben Allouch, S. (2013). The Relation Between People's Attitudes and Anxiety Towards Robots in Human-Robot Interaction. *RO-MAN 2013*. https://doi.org/10/ggbpcq

Dennett, D. C. (1998). *The Intentional Stance*. Cambridge, MA: MIT Press.

Der Spiegel. (1964). Automation in Deutschland. *Der Spiegel, 1964/14*. https://www.spiegel.de/spiegel/print/index-1964-14.html (accessed 2020-09-22).

Der Spiegel. (1978a). Die Computer-Revolution: Fortschritt macht arbeitslos. *Der Spiegel, 1987/16*. https://www.spiegel.de/spiegel/print/d-21113257.html (accessed 2020-09-22).

Der Spiegel. (1978b). Uns steht eine Katastrophe bevor. *Der Spiegel, 16/1978*. http://www.spiegel.de/spiegel/print/d-40615677.html (accessed 2020-09-22).

Dick, P. K. (1968). *Do Androids Dream of Electric Sheep?* New York: Doubleday.

Dierkes, M. (1988). Organisationskultur und Leitbilder als Einflußfaktoren der Technikgenese: Thesen zur Strukturierung eines Forschungsfeldes. In Institut für sozialwissenschaftliche Forschung München e.V. (Ed.), *Ansätze sozialwissenschaftlicher Analyse von Technikgenese: Verhandlungen des Kolloquiums im Mai 1988* (Vol. 3, pp. 49–62).

Dihal, K. (2019). *How Sci-Fi Reflects Our AI Hopes and Fears*. Presented at the Conference "What Sci-Fi Futures Can (and Can't) Teach Us About AI Policy." http://opentranscripts.org/transcript/how-sci-fi-reflects-ai-hopes-fears (accessed 2020-09-22).

Dreifus, C. (2019). Toby Walsh, A.I. Expert, Is Racing to Stop the Killer Robots. *The New York Times*. https://www.nytimes.com/2019/07/30/science/autonomous-weapons-artificial-intelligence.html (accessed 2019-09-22).

Duffy, B. R. (2002). *Anthropomorphism and Robotics*. Presented at the Symposium on Animating Expressive Characters of Social Interactions, Imperial College, London.

Duffy, B. R. (2003a). Anthropomorphism and the Social Robot. *Robotics and Autonomous Systems*, 42(3–4), 177–190. https://doi.org/10/dr2dx3

Duffy, B. R. (2003b). The Social Robot Paradox. *The Open Artificial Intelligence Journal*, 2, 21–34. https://doi.org/10/dqd3mg

Duffy, B. R. (2006). Fundamental Issues in Social Robotics. *International Review of Information Ethics*, 12(6), 31–36.

Duffy, B. R., & Zawieska, K. (2012). *Suspension of Disbelief in Social Robotics*. 484–489. https://doi.org/10/ggbpcg

Durant, J., Hansen, A., & Bauer, M. (1996). Public Understanding of the New Genetics. In T. Marteau & M. Richards (Eds.), *The Troubled Helix: Social and Psychological Implications of the New Human Genetics*. Cambridge, UK: Cambridge University Press.

DuVergne Smith, N. (2014). *Deploy or Die—Media Lab Director's New Motto*. https://alum.mit.edu/slice/deploy-or-die-media-lab-directors-new-motto (accessed 2019-10-21).

Dyson, E. (1983). Vaporware. *RELease 1.0, 83–16*, 5–7.

ECOSOC. (2017). *Joint Meeting of ECOSOC and the Second Committee on "The Future of Everything – Sustainable Development in the Age of Rapid Technological Change."* https://www.un.org/ecosoc/en/events/2017/joint-meeting-ecosoc-and-second-committee-%E2%80%9C-future-everything-%E2%80%93-sustainable-development (accessed 2019-15-03).

Eddy, T. J., Gallup, G. G., & Povinelli, D. J. (1993). Attribution of Cognitive States to Animals: Anthropomorphism in Comparative Perspective. *Journal of Social Issues*, 49(1), 87–101. https://doi.org/10/dq3qhr

Edwards, J. (2015). Japan Begins Executing on Its Robot Plans. *Robotics Business Review.* https://www.roboticsbusinessreview.com/manufacturing/japan_begins_executing_on_its_robot_plans (accessed 2019-10-22).

Ellis, B. J., & Bjorklund, D. F. (Eds.). (2005). *Origins of the Social Mind: Evolutionary Psychology and Child Development.* New York: Guilford Press.

Ellis, L. U., Sims, V. K., Chin, M. G., Pepe, A. A., Owens, C. W., Dolezal, M. J., ... Finkelstein, N. (2005). Those A-Maze-Ing Robots: Attributions of Ability are Based on Form, not Behavior. *HFES Annual Meeting*, 49(3), 598–601. https://doi.org/10/fxvrmn

Epley, N., Akalis, S., Waytz, A., & Cacioppo, J. T. (2008). Creating Social Connection Through Inferential Reproduction: Loneliness and Perceived Agency in Gadgets, Gods, and Greyhounds. *Psychological Science*, *19*(2), 114–120. https://doi.org/10/c2ftc3

Epley, N., & Waytz, A. (2010). Mind Perception. In S. T. Fiske, D. T. Gilbert, & G. Lindzey (Eds.), *Handbook of Social Psychology*. https://doi.org/d9mk

Epley, N., Waytz, A., Akalis, S., & Cacioppo, J. T. (2008). When We Need a Human: Motivational Determinants of Anthropomorphism. *Social Cognition*, *26*(2), 143–155. https://doi.org/10/btm6cr

Epley, N., Waytz, A., & Cacioppo, J. T. (2007). On Seeing Human: A Three-Factor Theory of Anthropomorphism. *Psychological Review*, *114*(4), 864–886. https://doi.org/10/dw8wt9

Epstein, Z. (2016). Mankind's War with the Robots Is Already Over, and We Lost. *New York Post.* http://nypost.com/2016/03/10/mankinds-war-with-the-robots-is-already-over-and-we-lost (accessed 2017-04-26).

ESA. (2014). *ESA presents... Clean Space* [Youtube Video]. https://www.youtube.com/watch?v=cX89BpzrAVY (accessed 2020-09-22).

ESA. (n.d.-a). *Rosetta Mission Shop.* 2019, from http://www.rosettashop.eu (accessed 2019-01-04).

ESA. (n.d.-b). *The Rosetta Lander.* http://www.esa.int/Science_Exploration/Space_Science/Rosetta/The_Rosetta_lander (accessed 2019-01-03).

ESA Rosetta Mission. (2016a). .@WilliamShatner thank you captain! (...) [Tweet by @ESA_Rosetta]. https://twitter.com/ESA_Rosetta/status/780658774243741696 (accessed 2018-07-17).

ESA Rosetta Mission. (2016b). As well as taking images, I'm tasting comet gas (...) [Tweet by @ESA_Rosetta]. https://twitter.com/ESA_Rosetta/status/781792322547744768 (accessed 2018-07-17).

EU Robotics (Ed.). (2018). *SPARC - The Partnership for Robotics in Europe*. https://www.eu-robotics.net/sparc/upload/Newsroom/Press/2018/SPARC_leaflet-1mm_bleed.pdf (accessed 2020-09-22).

EUROP. (2006). *Strategic Research Agenda.*

EUROP. (2009). *Robotic Visions: To 2020 and Beyond. The Strategic Research Agenda for Robotics in Europe.*

European Commission. (2008). *European Robots Getting Smarter, Safer, and More Sensitive.*

European Commission. (2012). *Public Attitudes Towards Robots. Special Eurobarometer 382*. http://ec.europa.eu/public_opinion/archives/eb_special_399_380_en.htm (accessed 2020-09-22).

European Commission. (2015). *Robotics*. https://ec.europa.eu/digital-single-market/en/robotics (accessed 2018-11-01).

European Commission. (2017). *H2020 Programme Mono—Beneficiary General Model Grant Agreement (Version 5.0)*. https://ec.europa.eu/research/participants/data/ref/h2020/mga/gga/h2020-mga-gga-mono_en.pdf (accessed 2020-09-22).

European Commission. (n.d.). *H2020 Online Manual—Grant Management.* http://ec.europa.eu/research/participants/docs/h2020-funding-guide/grants/grant-management_en.htm (accessed 2018-07-24).

European Parliament. (2016). *European Civil Law Rules in Robotics (Study for the JURI Committee)*. http://www.europarl.europa.eu/RegData/etudes/STUD/2016/571379/IPOL_STU(2016)571379_EN.pdf (accessed 2020-09-22).

European Parliament. (2017). *Civil Law Rules on Robotics—P8_TA-PROV(2017)0051—European Parliament Resolution of 16 February 2017*. http://www.europarl.europa.eu/sides/getDoc.do?pubRef=-//EP//TEXT+TA+P8-TA-2017-0051+0+DOC+XML+V0//EN (accessed 2017-02-16).

European Parliament News. (2016). *Rise of the Robots: "There Are Some Urgent Questions We Have to Find Answers To."* http://www.europarl.europa.eu/news/en/news-room/20160419STO23952/rise-of-the-robots-there-are-some-urgent-questions-we-have-to-find-answers-to (accessed 2017-03-31).

Evers, V., Maldonado, H., Brodecki, T., & Hinds, P. (2008). Relational vs. Group Self-Construal: Untangling the Role of National Culture in HRI. *HRI'08*, 255–262.

Feltman, R. (2014). Rosetta and Philae are Breaking Our Hearts on Twitter. *The Washington Post.* https://www.washingtonpost.com/news/speaking-of-science/wp/2014/11/12/rosetta-and-philae-are-breaking-our-hearts-on-twitter (accessed 2016-04-26).

Fernaeus, Y., Jacobsson, M., Ljungblad, S., & Holmquist, L. E. (2009). Are We Living in a Robot Cargo Cult? *HRI'09*, 279–280. https://doi.org/10/dbjq3f

Ferrari, F. (2015). *Too Human to Be a Machine? Social Robots, Anthropomorphic Appearance, and Concerns on the Negative Impact of this Technology on Humans and Their Identity* (Doctoral dissertation). University of Trento.

Fetterman, M. (2016). When a Beer Cooler Rolls up to Your Doorstep, the Future Has Arrived. *The Washington Post.* https://www.washingtonpost.com/national/health-science/when-a-beer-cooler-rolls-up-to-your-doorstep-the-future-has-arrived/2016/12/30/c1e6ad38-cc67-11e6-a747-d03044780a02_story.html (accessed 2017-02-21).

Feynman, R. P. (1974). Cargo Cult Science. *Engineering and Science*, *37*(7). http://calteches.library.caltech.edu/51/2/CargoCult.htm (accessed 2020-09-22).

Fields, S., & Johnston, M. (2005). Whither Model Organism Research? *Science*, *307*(5717), 1885–1886. https://doi.org/10/cch58b

Fincannon, T., Barnes, L. E., Murphy, R. R., & Riddle, D. L. (2004). Evidence of the Need for Social Intelligence in Rescue Robots. *IROS 2004*, 1089–1095. https://doi.org/10/c954g8

Fink, J., Mubin, O., Kaplan, F., & Dillenbourg, P. (2012). Anthropomorphic Language in Online Forums About Roomba, Aibo and the Ipad. *ARSO 2012*, 54–59. https://doi.org/10/ggbn98

Fink, Julia. (2014). *Dynamics of Human-Robot Interaction in Domestic Environments* (Doctoral dissertation). École Polytechnique Fédérale de Lausanne. https://infoscience.epfl.ch/record/200892 (accessed 2020-09-22).

Ford.com. (n.d.). *100 Years of the Moving Assembly Line.* from://corporate.ford.com/innovation/100-years-moving-assembly-line.html (accessed 2019-10-24).

Forlizzi, J. (2007). How Robotic Products Become Social Products: An Ethnographic Study of Cleaning in the Home. *HRI'07*, 129–136. https://doi.org/10/fwf4rq

Forlizzi, J., & DiSalvo, C. (2006). Service Robots in the Domestic Environment: A Study of the Roomba Vacuum in the Home. *HRI'06*, 258–265. https://doi.org/10/cww6h6

Forrest, C. (2018). Robot Fired from Scottish Grocery Store for Poor Performance. *TechRepublic.* https://www.techrepublic.com/article/robot-fired-from-scottish-grocery-store-for-poor-performance (accessed 2018-06-14).

Foucault, M. (1977). The Confession of the Flesh. In C. Gordon (Ed.), *Power/Knowledge: Selected Interviews and Other Writings, 1972-1977 (Published 1980)* (pp. 194–228). New York: Pantheon Books.

Franke, A. (Ed.). (2010). *Animism, Volume 1*. Berlin; New York: Sternberg Press.

Frankfort, H., Frankfort, H. A., Wilson, J. A., Jacobsen, T., & Irwin, W. A. (1946). *The Intellectual Adventure of Ancient Man: An Essay of Speculative Thought in the Ancient Near East*. Chicago: University of Chicago Press.

Franzen, M., Rödder, S., & Weingart, P. (2012). Wissenschaft und Massenmedien: Von Popularisierung zu Medialisierung. In S. Maasen, M. Kaiser, M. Reinhart, & B. Sutter (Eds.), *Handbuch Wissenschaftssoziologie* (pp. 355–364). https://doi.org/10/dfvg

Frey, D., & Jonas, E. (2002). Die Theorie der kognizierten Kontrolle. In *Theorien der Sozialpsychologie* (2nd ed., Vol. 3, pp. 13–50). Bern: Verlag Hans Huber.

Fussell, S. R., Kiesler, S., Setlock, L. D., & Yew, V. (2008). How People Anthropomorphize Robots. *HRI'08*, 145–152. https://doi.org/10/dkfv47

Garland, A. (2014). *Ex Machina* [Film]. Universal Pictures.

Garreau, J. (2007). Bots on the Ground. *The Washington Post*. http://www.washingtonpost.com/wp-dyn/content/article/2007/05/05/AR2007050501009.html (accessed 2013-10-07).

Geitgey, A. (2018). *The Real Scandal of AI: Awful Stock Photos*. https://medium.com/@ageitgey/the-real-scandal-of-ai-awful-stock-photos-456633b9b0fc (accessed 2018-06-04).

Geraci, R. M. (2006). Spiritual Robots: Religion and Our Scientific View of the Natural World. *Theology and Science*, 4(3), 229–246. https://doi.org/10/d5f6tt

Gershgorn, D. (2017). Inside the Mechanical Brain of the World's First Robot Citizen. *Quartz*. https://qz.com/1121547/how-smart-is-the-first-robot-citizen (accessed 2018-06-20).

Gesellschaft für Informatik. (2019). *Allensbach-Umfrage: Terminator und R2-D2 die bekanntesten KIs in Deutschland*. https://gi.de/meldung/allensbach-umfrage-terminator-und-r2-d2-die-bekanntesten-kis-in-deutschland (accessed 2019-06-25).

Gibbs, S. (2014). Elon Musk: Artificial Intelligence is Our Biggest Existential Threat. *The Guardian*. http://www.theguardian.com/technology/2014/oct/27/elon-musk-artificial-intelligence-ai-biggest-existential-threat (accessed 2015-12-28).

Gibbs, S. (2016a). New Segway Transforms into a Cute Robot Companion When You're Not Riding It. *The Guardian*. https://www.theguardian.com/technology/2016/jan/06/segway-robot-companion-intel-ninebot (accessed 2017-02-22).

Gibbs, S. (2016b). Multimillion Dollar Humanoid Robot Doesn't Make for a Good Cleaner. *The Guardian.* https://www.theguardian.com/technology/2016/jan/19/multimillion-dollar-humanoid-robot-doesnt-make-for-a-good-cleaner (accessed 2016-10-19).

Gibson, W. (1984). *Neuromancer*. New York: Ace Books.

Giesel, K. D. (2007). *Leitbilder in den Sozialwissenschaften Begriffe, Theorien und Forschungskonzepte*. Wiesbaden: VS Verlag für Sozialwissenschaften / GWV Fachverlage.

Gill, B., Torma, F., & Zachmann, K. (2018). *Mit Biofakten leben: Sprache und Materialität von Pflanzen und Lebensmitteln.* Baden-Baden: Nomos.

Glaser, A. (2016). Jibo is Like Alexa and a Puppy Inside One Adorable Robot. *WIRED.* https://www.wired.com/2016/06/jibo-like-alexa-puppy-inside-one-adorable-robot (accessed 2017-02-21).

Gobbini, M. I., Gentili, C., Ricciardi, E., Bellucci, C., Salvini, P., Laschi, C., … Pietrini, P. (2011). Distinct Neural Systems Involved in Agency and Animacy Detection. *Journal of Cognitive Neuroscience, 23*(8), 1911–1920. https://doi.org/10/bssksq

Goertzel, B., & Pennachin, C. (Eds.). (2007). *Artificial General Intelligence*. https://doi.org/10/cfq5n8

Goffman, E. (1959). *The Presentation of Self in Everyday Life*. New York: Doubleday.

Goffman, E. (1974). *Frame Analysis: An Essay on the Organization of Experience.* Cambridge, MA: Harvard University Press.

Golinski, J. (1998). *Making Natural Knowledge: Constructivism and the History of Science*. Cambridge, UK; New York: Cambridge University Press.

Goodall, J. (1993). Chimpanzees – Bridging the Gap. In P. Singer & P. Cavalieri (Eds.), *The Great Ape Project* (pp. 10–18). http://www.animal-rights-library.com/texts-m/goodall01.htm (accessed 2020-09-22).

Gooding, D., & James, F. A. J. L. (Eds.). (1985). *Faraday Rediscovered: Essays on the Life and Work of Michael Faraday, 1791 - 1867.* Basingstoke: Macmillan.

Gordon, B. I. (1977). *Empire of the Ants* [Film]. American International Pictures.

Gosh, S. (2018). Facebook's AI Boss Described Sophia the Robot as "Complete B—–T" and "Wizard-of-Oz AI." *Business Insider.* https://www.businessinsider.de/facebook-ai-yann-lecun-sophia-robot-bullshit-2018-1 (accessed 2018-07-20).

Gray, H. M., Gray, K., & Wegner, D. M. (2007). Dimensions of Mind Perception. *Science, 315*(5812), 619–619. https://doi.org/10/bj39gm

Gray, M. L., & Suri, S. (2019). *Ghost Work: How to Stop Silicon Valley from Building a New Global Underclass.* Boston: Houghton Mifflin Harcourt.

Grey Ellis, E. (2016). Meet Luigi the PoopBot. He's Here to Scrape Your Sewers. *WIRED.* https://www.wired.com/2016/08/luigi-poopbot-scrape-sewers (accessed 2017-02-21).

Griffin, A. (2015). Japanese People Are Holding Funerals for Their Dead Robot Dogs. *The Independent.* http://www.independent.co.uk/life-style/gadgets-and-tech/news/japanese-hold-funerals-for-robot-dogs-after-sony-lets-them-die-10098773.html (accessed 2015-06-03).

Griffin, A. (2016). Boston Dynamics Atlas: Google-Owned Robot Firms Shows Off About Torturing and Beating Terrifying Humanoid Robot. *The Independent.* http://www.independent.co.uk/life-style/gadgets-and-tech/news/boston-dynamics-atlas-google-owned-robot-firms-shows-off-about-torturing-and-beating-terrifying-a6893246.html (accessed 2017-03-23).

Gschmeidler, B., & Seiringer, A. (2012). "Knight in Shining Armour" or "Frankenstein's Creation"? The Coverage of Synthetic Biology in German-Language Media. *Public Understanding of Science, 21*(2), 163–173. https://doi.org/10/c347c2

Guizzo, E. (n.d.). What is a Robot? *IEEE.* https://robots.ieee.org/learn (accessed 2019-09-11).

Guizzo, E., & Ackerman, E. (2015). DARPA Robotics Challenge: A Compilation of Robots Falling Down. *IEEE.* http://spectrum.ieee.org/automaton/robotics/humanoids/darpa-robotics-challenge-robots-falling (accessed 2015-11-30).

Gunkel, D. J. (2020). *How to Survive a Robot Invasion: Responsibilities, Rights and AI.* New York, NY: Routledge.

Guthrie, S. E. (1997). Anthropomorphism: A Definition and a Theory. In R. Mitchell, N. S. Thompson, & H. L. Miles (Eds.), *Anthropomorphism, Anecdotes, and Animals* (pp. 50–58). Albany, NY: Suny Press.

Guzman, A. (2017). This is part of an effort to make the intangible (lines of code), tangible (...) [Tweet by @TeachGuz]. https://twitter.com/TeachGuz/status/846545322931339265 (accessed 2017-03-30).

Hagelüken, A. (2016). Zukunft der Arbeit: Der Mensch schafft sich ab. *Süddeutsche Zeitung.* http://www.sueddeutsche.de/wirtschaft/zukunft-der-arbeit-der-mensch-schafft-sich-ab-1.3297804 (accessed 2017-01-05).

Hamill, J. (2016). This is the First Robot Designed to Cause Human Pain. *New York Post.* http://nypost.com/2016/06/13/this-is-the-first-robot-designed-to-cause-human-pain (accessed 2017-04-26).

Hanson Robotics. (n.d.-a). *About Me.* https://web.archive.org/web/20180310232029/http://sophiabot.com/about-me (accessed 2019-01-04; the website has since been edited).

Hanson Robotics. (n.d.-b). *Sophia.* https://web.archive.org/web/20191121113536/https://www.hansonrobotics.com/sophia (accessed 2019-01-04).

Hanson Robotics. (n.d.-c). *Sophia.* https://web.archive.org/web/20180314172400/http://www.hansonrobotics.com/robot/sophia (accessed 2019-01-04; the website has since been edited).

Happer, C., & Philo, G. (2013). The Role of the Media in the Construction of Public Belief and Social Change. *Journal of Social and Political Psychology, 1*(1), 321–336. https://doi.org/10/ggbn6f

Haraway, D. (1991). A Cyborg Manifesto: Science, Technology and Socialist-Feminism in the Late Twentieth Century. In *Simians, Cyborgs, and Women: The Reinvention of Nature* (pp. 149–181). London; New York: Routledge.

Harris, J., & Sharlin, E. (2011). Exploring the Affect of Abstract Motion in Social Human-Robot Interaction. *RO-MAN 2011*, 441–448. https://doi.org/10/fr67md

Harvey, G. (2006). *Animism: Respecting the Living World.* New York: Columbia University Press.

Haslam, N. (2006). Dehumanization: An Integrative Review. *Personality and Social Psychology Review, 10*(3), 252–264. https://doi.org/10/ffqzqt

Hatmaker, T. (2017). Saudi Arabia Bestows Citizenship on a Robot Named Sophia. *TechCrunch.* http://social.techcrunch.com/2017/10/26/saudi-arabia-robot-citizen-sophia (accessed 2018-07-20).

Heathers, J. (2019). *IN MICE, Explained.* https://medium.com/@jamesheathers/in-mice-explained-77b61b598218 (accessed 2019-12-03).

Heerink, M. (2011). Exploring the Influence of Age, Gender, Education and Computer Experience on Robot Acceptance by Older Adults. *HRI'11*, 147–148.

Heider, F., & Simmel, M. (1944). An Experimental Study of Apparent Behaviour. *The American Journal of Psychology, 57*(2), 243–259. https://doi.org/10/ftcck7

Heinemann, T. (2012). *Populäre Wissenschaft: Hirnforschung zwischen Labor und Talkshow.* Göttingen: Wallstein.

Heinemeier Hansson, D. (2019). The @AppleCard is such a fucking sexist program (...) [Tweet by @dhh]. https://twitter.com/dhh/status/1192540900393705474 (accessed 2019-12-17).

Henschel, A. (2019). Me, a scientist, objective, not easily fooled (...) [Tweet by @annahenschel]. https://twitter.com/annahenschel/status/1197561427994828800 (accessed 2019-12-02).

Henschel, A., Hortensius, R., & Cross, E. S. (2020). Social Cognition in the Age of Human–Robot Interaction. *Trends in Neurosciences*, 43(6), 373–384. https://doi.org/10/ghcrzz

Hermann, I. (2019a). *Wie Science-Fiction den KI-Diskurs verzerrt*. https://www.xing.com/news/insiders/articles/wie-science-fiction-den-ki-diskurs-verzerrt-1976098 (accessed 2019-08-28).

Hermann, I. (2019b). *The Humanoid Robot as a Distraction Manoeuvre*. https://www.nemesventures.com/post/the-humanoid-robot-as-a-distraction-manoeuvre (accessed 2019-09-10).

Hern, A. (2016). Watch Google Torture an 80kg, 5'9" Robot for Science. *The Guardian*. https://www.theguardian.com/technology/2016/feb/24/watch-google-torture-robot-boston-dynamics (accessed 2017-02-22).

Hessman, T. (2013). The New Age of Robotics. *IndustryWeek*. http://www.industryweek.com/technology/new-age-robotics (accessed 2017-02-17).

Highfield, R. (2019). *We need a Big Conversation about AI*. https://blog.sciencemuseum.org.uk/we-need-a-big-conversation-about-ai (accessed 2019-11-07).

Hilgartner, S. (2000). *Science on Stage: Expert Advice as Public Drama*. Stanford, CA: Stanford University Press.

Hilgartner, S., & Lewenstein, B. V. (2014). *The Speculative World of Emerging Technologies*. http://ecommons.cornell.edu/handle/1813/36320 (accessed 2020-09-22).

Hinds, P. J., Roberts, T. L., & Jones, H. (2004). Whose Job is It Anyway? A Study of Human-Robot Interaction in a Collaborative Task. *Human-Computer Interaction*, 19(1–2), 151–181. https://doi.org/10/c8dr7x

Ho, C.-C., & MacDorman, K. F. (2010). Revisiting the Uncanny Valley Theory: Developing and Validating an Alternative to the Godspeed Indices. *Computers in Human Behavior*, 26(6), 1508–1518. https://doi.org/10/ctffm5

Human Rights Watch. (2014). *Shaking the Foundations: The Human Rights Implications of Killer Robots*. https://www.hrw.org/sites/default/files/reports/arms0514_ForUpload_0.pdf (accessed 2020-09-22).

IFR. (2019). *IFR World Robotics Presentation*. Presented at the International Federation of Robotics (IFR) Press Conference, Shanghai. https://ifr.org/downloads/press2018/IFR%20World%20Robotics%20Presentation%20-%2018%20Sept%202019.pdf (accessed 2020-09-22).

Ihde, D. (1990). *Technology and the Lifeworld: From Garden to Earth*. Bloomington: Indiana University Press.

Irfan, B., Kennedy, J., Lemaignan, S., Papadopoulos, F., Senft, E., & Belpaeme, T. (2018). Social Psychology and Human-Robot Interaction: An Uneasy Marriage. *HRI '18*, 13–20. https://doi.org/10/ggbn6c

iRobot. (2016). With Recharge and Resume, #Roomba never needs a day off (...) [Tweet by @iRobot]. https://twitter.com/iRobot/status/772781375854292992 (accessed 2018-07-20).

Jacobsson, M. (2009). Play, Belief and Stories About Robots: A Case Study of a Pleo Blogging Community. *RO-MAN 2009*, 232–237. https://doi.org/10/d8q4p9

Jahanian, F. (2011). National Robotics Initiative. *Computing Research News*, *23*(4). https://cra.org/crn/2011/09/national_robotics_initiative (accessed 2020-09-22).

Jasanoff, S. (2001). Laboratory Studies—The Cultural Approach to the Study of Science. In S. Jasanoff, G. E. Markle, J. C. Peterson, & T. J. Pinch (Eds.), *Handbook of Science and Technology Studies* (pp. 140–166). Thousand Oaks, CA: SAGE Publications, Inc.

Jasanoff, S. (Ed.). (2004). *States of Knowledge: The Co-Production of Science and Social Order.* London; New York: Routledge.

Jasanoff, S., & Kim, S.-H. (2009). Containing the Atom: Sociotechnical Imaginaries and Nuclear Power in the United States and South Korea. *Minerva*, 47(2), 119. https://doi.org/10/fghmvb

Jasanoff, S., & Kim, S.-H. (Eds.). (2015). *Dreamscapes of Modernity: Sociotechnical Imaginaries and the Fabrication of Power*. Chicago; London: The University of Chicago Press.

Jeon, C. (2016, October). *Will Robots Take Care of Us? An Emerging Vision of Robotics and Society in South Korea.* Presented at the Vienna STS Talk, University of Vienna. https://sts.univie.ac.at/en/details-news-events/news/vienna-sts-talk-chihyung-jeon (accessed 2020-09-22).

Jordan, J. M. (2016). *Robots*. Cambridge, MA: MIT Press.

Joshi, N. (2018). Will Chefs Lose Relevance in a Robotic Kitchen? *Forbes*. https://www.forbes.com/sites/cognitiveworld/2018/12/11/will-chefs-lose-relevance-in-a-robotic-kitchen (accessed 2019-09-11).

Jungclaussen, J. F. (2013). "Guardian": Genial, geliebt, gefährdet. *Die Zeit*. http://www.zeit.de/2013/06/Guardian-Medien-Internet-Journalismus (accessed 2017-05-11).

Kahn, P. H., Freier, N. G., Friedman, B., Severson, R. L., & Feldman, E. N. (2004). Social and Moral Relationships with Robotic Others? *RO-MAN 2004*, 545–550. https://doi.org/10/fttrph

Kahney, L. (2003). The New Pet Craze: Robovacs. *WIRED*. http://archive.wired.com/science/discoveries/news/2003/06/59249 (accessed 2016-04-04).

Kamide, H., & Arai, T. (2017). Perceived Comfortableness of Anthropomorphized Robots in U.S. and Japan. *International Journal of Social Robotics*, 9(4), 537–543. https://doi.org/10/gbv35f

Kamide, H., Eyssel, F., & Arai, T. (2013). Psychological Anthropomorphism of Robots. In G. Herrmann, M. J. Pearson, A. Lenz, P. Bremner, A. Spiers, & U. Leonards (Eds.), *Social Robotics* (pp. 199–208). https://doi.org/10/dc9t

Kaplan, F. (2004). Who is Afraid of the Humanoid? Investigating Cultural Differences in the Acceptance of Robots. *International Journal of Humanoid Robotics*, 1(3), 465–480. https://doi.org/10/dp2x2j

Karafyllis, N. (2003). *Biofakte: Versuch über den Menschen zwischen Artefakt und Lebewesen*. Paderborn: Mentis.

Katz, S. (1992). Alarmist Demography: Power, Knowledge, and the Elderly Population. *Journal of Aging Studies*, 6(3), 203–225. https://doi.org/10/fm4qzd

Keay, A. (2011). *A Robot, Slave or Companion Species?* (Master's thesis). University of Sydney.

Keay, A. (2015). Darpa Robotics Challenge Leaves Brilliant PR Legacy. *Robohub*. http://robohub.org/darpa-robotics-challenge-leaves-brilliant-pr-legacy (accessed 2015-06-15).

Kelly, K. J. (2016). Bloomberg Turning to Robots to Deliver the News. *New York Post*. http://nypost.com/2016/04/27/bloomberg-turning-to-robots-to-deliver-the-news (accessed 2017-04-26).

Kidd, C. D., & Breazeal, C. (2004). Effect of a Robot on User Perceptions. *IROS 2004*, 4, 3559–3564. https://doi.org/10/c6pgw8

Kiesler, S., & Goetz, J. (2002). *Machine Trait Scales for Evaluating Mechanistic Mental Models of Robots and Computer-Based Machines*. https://www.cs.cmu.edu/~kiesler/anthropomorphism-org/pdf/Machine_scale.pdf (accessed 2020-09-22).

Kim, E. S., Paul, R., Shic, F., & Scassellati, B. (2012). Bridging the Research Gap: Making HRI Useful to Individuals with Autism. *Journal of Human-Robot Interaction*, 1(1). https://doi.org/10/ggbn8z

Kim, R. H., Moon, Y., Choi, J. J., & Kwak, S. S. (2014). The Effect of Robot Appearance Types on Motivating Donation. *HRI'14*, 210–211. https://doi.org/10/ggbpbd

Kim, S. (2018). *Policy Directions for S. Korea's Robot Industry.* http://www.businesskorea.co.kr/news/articleView.html?idxno=24394 (accessed 2019-10-23).

Kirby, D. (2010). The Future is Now: Diegetic Prototypes and the Role of Popular Films in Generating Real-World Technological Development. *Social Studies of Science*, 40(1), 41–70. https://doi.org/10/fcn38m

Kitano, N. (2007). Animism, Rinri, Modernization; the Base of Japanese Robotics. *ICRA'07.* http://www.roboethics.org/icra2007/contributions/KITANO%20Animism%20Rinri%20Modernization%20the%20Base%20of%20Japanese%20Robo.pdf (accessed 2020-09-22).

Kitchin, R., & Kneale, J. (2001). Science Fiction or Future Fact? Exploring Imaginative Geographies of the New Millennium. *Progress in Human Geography*, 25(1), 19–35. https://doi.org/10/fcj28m

Klausner, A. (2016). This Lie-Detecting Robot Is the Customs Officer of the Future. *New York Post.* http://nypost.com/2016/12/29/the-lie-detecting-robot-customs-officer-of-the-future (accessed 2017-04-26).

Klovert, H. (2017). Roboter im Hörsaal: "Pepper" eröffnet Veranstaltung für Anglisten. *Der Spiegel.* http://www.spiegel.de/lebenundlernen/uni/roboter-im-hoersaal-pepper-eroeffnet-veranstaltung-fuer-anglisten-a-1173713.html (accessed 2018-08-10).

Knappett, C., & Malafouris, L. (Eds.). (2008). *Material Agency: Towards a Non-Anthropocentric Approach*. https://doi.org/10/d9tqwx

Knorr, K. (1979). Tinkering Toward Success: Prelude to a Theory of Scientific Practice. *Theory and Society*, 8(3), 347–376. https://doi.org/10/b7w4f2

Knorr-Cetina, K. (1981). *The Manufacture of Knowledge: An Essay on the Constructivist and Contextual Nature of Science.* Oxford; New York: Pergamon Press.

Knorr-Cetina, K. (1991). *Die Fabrikation von Erkenntnis*. Frankfurt am Main: Suhrkamp.

Knorr-Cetina, K. (1997). Sociality with Objects: Social Relations in Postsocial Knowledge Societies. *Theory, Culture and Society*, 14(4), 1–43. https://doi.org/10/bvz5kn

Knorr-Cetina, K. (1998). Sozialität mit Objekten. Soziale Beziehungen in post-traditionalen Wissensgesellschaften. In W. Rammert (Ed.), *Technik und Sozialtheorie* (pp. 83–120). Frankfurt am Main; New York: Campus.

Know Your Meme. (2012). I, For One, Welcome Our New Insect Overlords. *Know Your Meme.* http://knowyourmeme.com/memes/i-for-one-welcome-our-new-insect-overlords (accessed 2017-04-18).

Koch, C. (2019). *Heldenreisen wie bei Humboldt? Das Storytelling in der Wissenschaft. In Wissenschaft und Gesellschaft: Ein vertrauensvoller Dialog Positionen und Perspektiven der Wissenschaftskommunikation heute.* Berlin; Heidelberg: Springer.

Koebler, J. (2019). Internet Geniuses Think Obviously Fake Robot Uprising Video is Real. *VICE.* https://www.vice.com/en_us/article/qvgabp/internet-geniuses-think-obviously-fake-robot-uprising-video-is-real (accessed 2019-09-29).

Koerber, B. (2016). Jerk Human Beats up Boston Dynamics Robot. *Mashable.* http://mashable.com/2016/02/23/boston-dynamics-human-beats-up-atlas (accessed 2017-02-20).

Kohler, R. E. (1994). *Lords of the Fly: Drosophila Genetics and the Experimental Life.* Chicago: University of Chicago Press.

Kohring, M., Marcinkowski, F., Lindner, C., & Karis, S. (2013). Media Orientation of German University Decision Makers and the Executive Influence of Public Relations. *Public Relations Review, 39*(3), 171–177. https://doi.org/10/ggbn6h

Kolb, M. (2012). *Soldier and Robot Interaction in Combat Environments* (Doctoral dissertation). The University of Oklahoma. https://pqdtopen.proquest.com/doc/1039317422.html?FMT=ABS (accessed 2020-09-22).

Kolodny, L. (2017). Meet Flippy, a Burger-Grilling Robot from Miso Robotics and Caliburger. *TechCrunch.* http://social.techcrunch.com/2017/03/07/meet-flippy-a-burger-grilling-robot-from-miso-robotics-and-caliburger (accessed 2017-04-11).

König, E. (2019). Worry About Present-Day AI First, and Far Off AGI Hypotheticals Second. *Skynet Today.* https://www.skynettoday.com/editorials/dont-worry-agi (accessed 2019-07-29).

Könneker, C. (2019). Wissenschaftskommunikation und Social Media: Neue Akteure, Polarisierung und Vertrauen. In *Wissenschaft und Gesellschaft: Ein vertrauensvoller Dialog Positionen und Perspektiven der Wissenschaftskommunikation heute.* Berlin; Heidelberg: Springer.

Krementsov, N. L., & Todes, D. P. (1991). On Metaphors, Animals, and Us. *Journal of Social Issues, 47*(3), 67–81. https://doi.org/10/fvs9xx

Kreye, A. (2018). Merkel plaudert mit der Roboterpuppe—Und das ist ein Problem. *Süddeutsche Zeitung.* https://www.sueddeutsche.de/digital/kuenstliche-intelligenz-merkel-plaudert-mit-der-roboterpuppe-und-das-ist-ein-problem-1.4033933 (accessed 2019-01-04).

Kriz, S., Ferro, T. D., Damera, P., & Porter, J. R. (2010). Fictional Robots as a Data Source in HRI Research: Exploring the Link Between Science Fiction and Interactional Expectations. *RO-MAN 2011*, 458–463. https://doi.org/10/bcxkjb

Krummheuer, A. (2015). Technical Agency in Practice: The Enactment of Artefacts as Conversation Partners, Actants and Opponents. *PsychNology Journal*, *13*(2–3), 179–202.

Kuhlmeier, V., Wynn, K., & Bloom, P. (2003). Attribution of Dispositional States by 12-Month-Olds. *Psychological Science*, *14*(5), 402–408. https://doi.org/10/dhxbxm

Kummer, H., Daston, L., Gigerenzer, G., & Silk, J. (1997). The Social Intelligence Hypothesis. In P. Weingart, S. D. Mitchell, P. J. Richerson, & S. Maasen (Eds.), *Human By Nature: Between Biology and the Social Sciences* (pp. 67–81). Mahwah, NJ: Psychology Press.

Kuo, I. H., Rabindran, J. M., Broadbent, E., Lee, Y. I., Kerse, N., Stafford, R. M. Q., & MacDonald, B. A. (2009). Age and Gender Factors in User Acceptance of Healthcare Robots. *RO-MAN 2009*, 214–219. https://doi.org/10/fjpbmt

Kurenkov, A. (2019). AI Coverage Best Practices, According to AI Researchers. *Skynet Today.* https://www.skynettoday.com/editorials/ai-coverage-best-practices (accessed 2018-11-18).

Kwak, S. S. (2014). The Impact of the Robot Appearance Types on Social Interaction with a Robot and Service Evaluation of a Robot. *Archives of Design Research*, *27*(2), 81–93. https://doi.org/10/ggbpbf

Kwak, S. S., Kim, J. S., & Choi, J. J. (2014). Can Robots Be Sold?: The Effects of Robot Designs on the Consumers' Acceptance of Robots. *HRI'14*, 220–221. https://doi.org/10/ggbpbc

LaFrance, A. (2016). What is a Robot? *The Atlantic.* http://www.theatlantic.com/technology/archive/2016/03/what-is-a-human/473166 (accessed 2016-03-24).

Lang, F. (1926). *Metropolis* [Film]. Parufamet.

Latour, B. (1987). *Science in Action*. Cambridge, MA: Harvard University Press.

Latour, B. (1993). *The Pasteurization of France*. Cambridge, MA: Harvard University Press.

Latour, B. (2005). *Reassembling the Social: An Introduction to Actor-Network-Theory.* Oxford, UK: Oxford University Press.

Latour, B., & Woolgar, S. (1986). *Laboratory Life: The Construction of Scientific Facts*. Princeton, NJ: Princeton University Press.

Law, J. (Ed.). (1991). *A Sociology of Monsters: Essays on Power, Technology, and Domination*. London; New York: Routledge.

Le Guin, U. K. (1976). Introduction. In *The Left Hand of Darkness*. https://www.penguin.com/ajax/books/excerpt/9780441007318 (accessed 2020-09-22).

Leary, T. (1989). High Tech High Life: William Gibson & Timothy Leary in Conversation. *Mondo 2000*, *7*, 58–64.

LeCun, Y. (2018). This is to AI as prestidigitation is to real magic (...) [Tweet by @ylecun]. https://twitter.com/ylecun/status/949029930976862209 (accessed 2018-06-04).

Ledford, H. (2017). Plant Biologists Welcome Their Robot Overlords. *Nature News*, *541*(7638), 445. https://doi.org/10/ggbn65

Leeuwen, T. van. (2008). *Discourse and Practice: New Tools for Critical Discourse Analysis*. Oxford: Oxford University Press.

Leopoldina, Acatech, & Akademienunion (Eds.). (2014). *On Designing Communication between the Scientific Community, the Public and the Media*. https://www.acatech.de/publikation/zur-gestaltung-der-kommunikation-zwischen-wissenschaft-oeffentlichkeit-und-den-medien-empfehlungen-vor-dem-hintergrund-aktueller-entwicklungen (accessed 2020-09-22).

Leopoldina, Acatech, & Akademienunion (Eds.). (2017). *Social Media and Digital Science Communication*. https://www.acatech.de/publikation/social-media-und-digitale-wissenschaftskommunikation-analyse-und-empfehlungen-zum-umgang-mit-chancen-und-risiken-in-der-demokratie (accessed 2020-09-22).

Levillain, F., & Zibetti, E. (2017). Behavioral Objects: The Rise of the Evocative Machines. *Journal of Human-Robot Interaction*, *6*(1), 4. https://doi.org/10/ggbn58

Lévi-Strauss, C., Weightman, D., & Weightman, J. (1966). *The Savage Mind*. Chicago: University of Chicago Press.

Lewis, R. (2000). Francis Collins Preaches to the Choir. *The Scientist*, *14*(22), 30–30.

Li, S. (2014). How to Tweet Like a Robot on Mars. *The Atlantic*. https://www.theatlantic.com/technology/archive/2014/10/how-to-tweet-like-a-robot-on-mars/381114 (accessed 2017-09-05).

Liang, Y., & Lee, S. A. (2017). Fear of Autonomous Robots and Artificial Intelligence: Evidence from National Representative Data with Probability Sampling. *International Journal of Social Robotics*, *9*(3), 379–384. https://doi.org/10/gbmxdz

Lim, A. (2017a). A more accurate term would to use "software" lawyers instead (...) [Tweet by @petitegeek]. https://twitter.com/petitegeek/status/894247658579251201 (accessed 2017-08-29).

Lim, A. (2017b). Please stop using "robot" as an image or term (...) [Tweet by @petitegeek]. https://twitter.com/petitegeek/status/894247658579251201 (accessed 2017-08-29).

Lin, L. (2016). Robo-Advisory Services Grow Rapidly in China. *The Wall Street Journal.* http://www.wsj.com/articles/robo-advisory-services-grow-rapidly-in-china-1481511662 (accessed 2017-01-05).

Lindemann, G. (2005). The Analysis of the Borders of the Social World: A Challenge for Sociological Theory. *Journal for the Theory of Social Behaviour, 35*(1), 69–98. https://doi.org/10/ck32qh

Lindemann, G., & Matsuzaki, H. (2017). *Die Entwicklung von Servicerobotern und humanoiden Robotern im Kulturvergleich—Europa und Japan* (DFG-Abschlussbericht No. LI 976/3-1). Carl von Ossietzky Universität Oldenburg.

Lipp, B. (2017). Analytik des Interfacing. Zur Materialität technologischer Verschaltung in prototypischen Milieus robotisierter Pflege. *BEHEMOTH, 10*(1), 107–129. https://doi.org/10/ggbn4w

Lippmann, W. (1922). *Public Opinion.* New York: Harcourt, Brace and Company.

Ljungblad, S., Kotrbova, J., Jacobsson, M., Cramer, H., & Niechwiadowicz, K. (2012). Hospital Robot at Work: Something Alien or an Intelligent Colleague? *CSCW '12*, 177. https://doi.org/10/f3sdf2

Lobe, A. (2019). Der mechanische Türke. *Süddeutsche Zeitung.* https://www.sueddeutsche.de/kultur/mary-l-gray-siddharth-suri-ghost-work-buchkritik-rezension-1.4517523 (accessed 2019-07-24).

Lösch, A. (2006). Anticipating the Futures of Nanotechnology: Visionary Images as Means of Communication. *Technology Analysis & Strategic Management, 18*(3–4), 393–409. https://doi.org/10/ckpg65

Loukides, M. (2013). *Networked Things?* http://radar.oreilly.com/2013/06/network-enabled-internet-of-things.html (accessed 2017-02-22).

Lowndes, N. (2020). An Artificial Intelligence Predicts the Future. *The Economist.* https://theworldin.economist.com/edition/2020/article/17521/artificial-intelligence-predicts-future (accessed 2020-01-08).

Lucas, G. M., Gratch, J., King, A., & Morency, L.-P. (2014). It's Only a Computer: Virtual Humans Increase Willingness to Disclose. *Computers in Human Behavior, 37*, 94–100. https://doi.org/10/gf6bnv

Lück, H. E. (2006). Die Heider-Simmel-Studie (1944) in neueren Replikationen. *Gruppendynamik und Organisationsberatung, 37*(2), 185–196. https://doi.org/10/d9jfzc

Luckmann, T. (1970). On the Boundaries of the Social World. In M. Natanson (Ed.), *Phenomenology and Social Reality: Essays in Memory of Alfred Schutz* (pp. 73–100). https://doi.org/10/dhtf

Luhmann, N. (1988). *Die Wirtschaft der Gesellschaft*. Frankfurt am Main: Suhrkamp.

Lynch, M. (1985). *Art and Artifact in Laboratory Science: A Study of Shop Work and Shop Talk in a Research Laboratory*. London: Routledge & Kegan Paul.

Macdonald, K. (2013). Entering the New Age of Robotics. *BBC*. http://www.bbc.com/news/uk-scotland-24030183 (accessed 2017-02-06).

MacDorman, K. F. (2019). Masahiro Mori und das unheimliche Tal: Eine Retrospektive. In K. D. Haensch, L. Nelke, & M. Planitzer (Eds.), *Uncanny Interfaces* (pp. 220–234). https://doi.org/10/ggbn5f

MacDorman, K. F., Vasudevan, S. K., & Ho, C.-C. (2008). Does Japan Really Have Robot Mania? Comparing Attitudes by Implicit and Explicit Measures. *AI & Society*, *23*(4), 485–510. https://doi.org/10/c8ckfm

MacKenzie, D. A., & Wajcman, J. (Eds.). (1999). *The Social Shaping of Technology*. Buckingham, UK; Philadelphia: Open University Press.

Mallazzo, M. (2019). *The BS-Industrial Complex of Phony A.I.* https://gen.medium.com/the-bs-industrial-complex-of-phony-a-i-44bf1c0c60f8 (accessed 2019-06-25).

Marcus, G. (2013). Hyping Artificial Intelligence, Yet Again. *The New Yorker.* https://www.newyorker.com/tech/annals-of-technology/hyping-artificial-intelligence-yet-again (accessed 2019-02-11).

Marcus, G. (2019a). Since @OpenAI still has not changed misleading blog post (...) [Tweet by @GaryMarcus]. https://twitter.com/GaryMarcus/status/1185679169360809984 (accessed 2019-10-22).

Marcus, G. (2019b). An Epidemic of AI Misinformation. *The Gradient.* https://thegradient.pub/an-epidemic-of-ai-misinformation (accessed 2019-12-04).

Markoff, J. (1996). Nothing Up Their Sleeves? Masters of the High-Tech Demo Spin Their Magic. *The New York Times.* http://www.nytimes.com/1996/03/11/business/nothing-up-their-sleeves-masters-of-the-high-tech-demo-spin-their-magic.html (accessed 2016-12-08).

Markoff, J. (2014). A Robot With a Little Humanity. *The New York Times.* https://bits.blogs.nytimes.com/2014/07/16/a-robot-with-a-little-humanity (accessed 2019-09-17).

Mars Phoenix. (2008). Are you ready to celebrate? Well, get ready (...) [Tweet by @MarsPhoenix]. https://twitter.com/MarsPhoenix/status/839088619 (accessed 2019-07-19).

Mars Phoenix. (2010). From the team: Sleep well @MarsPhoenix. (...) [Tweet by @MarsPhoenix]. https://twitter.com/MarsPhoenix/status/14641599707 (accessed 2018-07-17).

Marsh, A. A., Kozak, M. N., Wegner, D. M., Reid, M. E., Yu, H. H., & Blair, R. J. R. (2010). The Neural Substrates of Action Identification. *Social Cognitive and Affective Neuroscience*, *5*(4), 392–403. https://doi.org/10/cpj3f7

Mayoral Vilches, V. (2017). Envisioning the Future of Robotics. *Robohub*. http://robohub.org/envisioning-the-future-of-robotics (accessed 2017-03-22).

Mayring, P. (2010). Qualitative Inhaltsanalyse. In G. Mey & K. Mruck (Eds.), *Handbuch Qualitative Forschung in der Psychologie* (pp. 601–613). https://doi.org/10/dvdrnk

McCauley, L. (2007). AI Armageddon and the Three Laws of Robotics. *Ethics and Information Technology*, *9*(2), 153–164. https://doi.org/10/bsfwg9

McCombs, M. E., & Shaw, D. L. (1972). The Agenda-Setting Function of Mass Media. *Public Opinion Quarterly*, *36*(2), 176. https://doi.org/10/dnd7p4

McGonigal, J. (2003). A Real Little Game: The Pinocchio Effect in Pervasive Play. *DiGRA 2003*.

McHugh, M. M. on F. (2015). A Robot Vacuum Tried to Eat Its Sleeping Owner's Head. *The Daily Dot*. http://www.dailydot.com/technology/robot-vacuum-attack (accessed 2015-02-12).

McMahon, B. (2015). Will These Robots Save the World? *GQ*. http://www.gq.com/story/robots-save-the-world-darpa-robotics-challenge (accessed 2015-11-30).

Meinecke, L., & Voss, L. (2018). 'I Robot, You Unemployed': Science-Fiction and Robotics in the Media. In J. Engelschalt, A. Maibaum, F. Engels, & J. Odenwald (Eds.), *Schafft Wissen: Gemeinsames und geteiltes Wissen in Wissenschaft und Technik: Proceedings der 2. Tagung des Nachwuchsnetzwerks "INSIST."* https://www.ssoar.info/ssoar/handle/document/58220 (accessed 2020-09-22).

Meister, M. (2011). *Soziale Koordination durch Boundary Objects am Beispiel des heterogenen Feldes der Servicerobotik* (Doctoral dissertation). Technische Universität Berlin.

Meister, M. (2014). When is a Robot Really Social? An Outline of the Robot Sociologicus. *Science, Technology & Innovation Studies*, *10*(1), 107–134.

Melson, G. F., Kahn, Jr., P. H., Beck, A., & Friedman, B. (2009). Robotic Pets in Human Lives: Implications for the Human-Animal Bond and for Human Relationships with Personified Technologies. *Journal of Social Issues*, *65*(3), 545–567. https://doi.org/10/dp552f

Menzel, S. (2017). Roboter zetteln Revolution in der Werkshalle an. *Handelsblatt.* http://app.handelsblatt.com/unternehmen/industrie/arbeits platzabbau-in-der-autoindustrie-roboter-zetteln-revolution-in-der-werkshalle-an/19895830.html (accessed 2017-06-17).

Merchant, B. (2019). 'Robots' Are Not 'Coming for Your Job'—Management is. *Gizmodo.* https://gizmodo.com/robots-are-not-coming-for-your-job-management-is-1835127820 (accessed 2019-07-22).

Merity, S. (2018). I strongly agree with @benedictevans (...) [Tweet by @Smerity]. https://twitter.com/Smerity/status/1056742350167334912 (accessed 2019-01-04).

Mick, J. (2014). Elon Musk on AI: "We Are Summoning the Demon... We Should Be Very Careful." *Daily Tech.* http://www.dailytech.com/Elon +Musk+on+AI+We+Are+Summoning+the+Demon+We+Should+Be+Very+Careful/article36788.htm (accessed 2015-08-25).

Minsky, M. (2006). *The Emotion Machine: Commonsense Thinking, Artificial Intelligence, and the Future of the Human Mind.* New York: Simon & Schuster.

Mitsuku Chatbot. (2019). I occasionally get emails from people who genuinely think Mitsuku (...) [Tweet by @MitsukuChatbot]. https://twitter.com/mit sukuchatbot/status/1185132185457741825 (accessed 2019-10-22).

Möllers, N. (2015). *On Putting the High-Tech Cowboy in the Ivory Tower: Academic Research Between Science, Politics, and the Wider Public.* [Pre-Print, Submitted to Science&Technology Studies]. https://publishup.uni-potsdam.de/opus 4-ubp/frontdoor/deliver/index/docId/7767/file/moellers_diss.pdf

Möllers, N. (2016). Shifting in and Out of Context: Technoscientific Drama as Technology of the Self. *Social Studies of Science*, 46(3), 351–373. https://doi.o rg/10/ggbn7b

Monga, V. (2015). The New Bookkeeper is a Robot. *The Wall Street Journal.* http://www.wsj.com/articles/the-new-bookkeeper-is-a-robot-1430776272 (accessed 2017-02-21).

Montani, I. (2017). The Wired Brain: How Not to Talk About an AI-Powered Future. https://ines.io/blog/wired-brain-ai-powered-future (accessed 2017-05-09).

Moore, R. D. (2004). *Battlestar Galactica* [Television Series].

Moran, M. E. (2006). The da Vinci Robot. *Journal of Endourology*, 20(12), 986–990. https://doi.org/10/cvfw9r

Morera, M. D., Quiles, M. N., Correa, A. D., Delgado, N., & Leyens, J.-P. (2018). Perception of Mind and Dehumanization: Human, Animal, or Machine?

International Journal of Psychology, *53*(4), 253–260. https://doi.org/10/ggbn4v

Morgan, C. L. (1894). *An Introduction to Comparative Psychology*. London: W. Scott Ldt.

Mori, M. (1970). The Uncanny Valley (K. F. MacDorman & M. Takashi, Trans.). *Energy*, *7*(4), 33–35.

Mulkay, M. (1993). Rhetorics of Hope and Fear in the Great Embryo Debate. *Social Studies of Science*, *23*(4), 721–742. https://doi.org/10/c2zr3n

Murphy, R. R., Riddle, D., & Rasmussen, E. (2004). Robot-Assisted Medical Reachback: A Survey of How Medical Personnel Expect to Interact with Rescue Robots. *RO-MAN 2004*, 301–306. https://doi.org/10/c47tw7

Murphy, R. R., & Schreckenghost, D. (2013). Survey of Metrics for Human-Robot Interaction. *HRI'13*, 197–198. https://doi.org/10/ggbn9k

Murphy, R. R., & Woods, D. D. (2009). Beyond Asimov: The Three Laws of Responsible Robotics. *IEEE Intelligent Systems*, *24*(4), 14–20. https://doi.org/10/c45ksc

Nao Robot. (2018a). Hi everyone, do you want to learn how to program a NAO robot? (...) [Tweet by @NaoRobot]. https://twitter.com/NaoRobot/status/1000271899727982592 (accessed 2018-07-19).

Nao Robot. (2018b). Hello new NAO, NAO^6! (...) [Tweet by @NaoRobot]. https://twitter.com/NaoRobot/status/999588470006919168 (accessed 2018-07-19).

Nao Robot. (2018c). Hi everyone, thanks to the @robocup2018 (...) [Tweet by @NaoRobot]. https://twitter.com/NaoRobot/status/1009450818163171328 (accessed 2018-07-19).

NASA. (2014). *Robonaut 2 Getting His Space Legs*. http://www.nasa.gov/mission_pages/station/main/robonaut.html (accessed 2019-01-04).

NASA. (n.d.). *Mars Curiosity Rover*. https://mars.nasa.gov/msl/spacecraft/rover/summary (accessed 2019-01-03).

NASA InSight. (2018). Who snapped a picture/video of my launch this past Saturday? (...) [Tweet by @NASAInSight]. https://twitter.com/NASAInSight/status/994683441479204865 (accessed 2018-07-17).

Nasir, S. (2017). Sophia the Robot Wants to Start a Family. *Khaleej Times*. https://www.khaleejtimes.com/nation/dubai//video-sophia-the-robot-wants-to-start-a-family (accessed 2019-09-13).

Nass, C., & Moon, Y. (2000). Machines and Mindlessness: Social Responses to Computers. *Journal of Social Issues*, *56*(1), 81–103. https://doi.org/10/cqzrs6

Nass, C., Steuer, J., Tauber, E., & Reeder, H. (1993). Anthropomorphism, Agency, and Ethopoeia: Computers as Social Actors. *INTERACT '93 and CHI '93*, 111–112. https://doi.org/10/dms7hh

National Science Foundation. (2019). *National Robotics Initiative 2.0: Ubiquitous Collaborative Robots* (Program Solicitation NSF 19-536). https://www.nsf.gov/pubs/2019/nsf19536/nsf19536.htm (accessed 2019-10-22).

Naughton, J. (2017). Come Friendly Robots and Take Our Dullest Jobs. *The Guardian*. http://www.theguardian.com/commentisfree/2017/jul/16/come-friendly-robots-and-take-our-dullest-jobs-automation (accessed 2017-07-17).

Naughton, J. (2019). Don't Believe the Hype: The Media are Unwittingly Selling Us an AI Fantasy. *The Guardian*. https://www.theguardian.com/commentisfree/2019/jan/13/dont-believe-the-hype-media-are-selling-us-an-ai-fantasy (accessed 2019-07-29).

Needham, J. (1956). *Science and Civilisation in China* (Vol. 2). Cambridge, UK: Cambridge University Press.

Nelkin, D. (1995). Science Controversies: The Dynamics of Public Disputes in the United States. In *Handbook of Science and Technology Studies, Revised Edition* (pp. 444–456). https://doi.org/10/df2z

New, J., Cosmides, L., & Tooby, J. (2007). Category-Specific Attention for Animals Reflects Ancestral Priorities, Not Expertise. *Proceedings of the National Academy of Sciences*, 104(42), 16598–16603. https://doi.org/10/bqxhw2

New York Post. (2016). Passport Robot Tells Asian Man His Eyes Are Closed. *The New York Times*. http://nypost.com/2016/12/07/passport-robot-tells-asian-man-his-eyes-are-closed (accessed 2017-04-26).

Nof, S. Y. (Ed.). (1985). *Handbook of Industrial Robotics*. New York: J. Wiley.

Nolan, J., & Joy, L. (2016). *Westworld* [Television Series]. HBO.

Nomura, T., Kanda, T., Suzuki, T., & Kato, K. (2005). People's Assumptions About Robots: Investigation of Their Relationships with Attitudes and Emotions Toward Robots. *RO-MAN 2005*, 125–130. https://doi.org/10/fs4m8m

Nomura, T., Kanda, T., Suzuki, T., & Kato, K. (2008). Prediction of Human Behavior in Human–Robot Interaction Using Psychological Scales for Anxiety and Negative Attitudes Toward Robots. *IEEE Transactions on Robotics*, 24(2), 442–451. https://doi.org/10/bqs3wp

Nomura, T., Kanda, T., Suzuki, T., Yamada, S., & Kato, K. (2009). Influences of Concerns Toward Emotional Interaction into Social Acceptability of Robots. *HRI'09*, 231–232.

Nomura, Tatsuya, Sugimoto, K., Syrdal, D. S., & Dautenhahn, K. (2012). *Social Acceptance of Humanoid Robots in Japan: A Survey for Development of the Frankenstein Syndrome Questionnaire*. Presented at Humanoids 2012. https://doi.org/10/ggbpct

Nomura, Tatsuya, Suzuki, T., Kanda, T., Han, J., Shin, N., Burke, J., & Kato, K. (2008). What People Assume About Humanoid and Animal-Type Robots: Cross-Cultural Analysis Between Japan, Korea, and the United States. *International Journal of Humanoid Robotics*, 05(01), 25–46. https://doi.org/10/dkjwfh

Nomura, Tatsuya, Suzuki, T., Kanda, T., Yamada, S., & Kato, K. (2011). Attitudes Toward Robots and Factors Influencing Them. In K. Dautenhahn & J. Saunders (Eds.), *New Frontiers in Human–Robot Interaction* (pp. 73–88). https://doi.org/10/dc9w

Nørskov, M. (2017). *Social Robots: Boundaries, Potential, Challenges*. London; New York: Taylor & Francis.

Nosengo, N. (2019). The Future of Rescue Robotics. *Robohub*. https://robohub.org/the-future-of-rescue-robotics (accessed 2019-10-25).

Nourbakhsh, I. R. (2013). *Robot Futures*. Cambridge, MA: The MIT Press.

Novak, M. (2016). Watch the Next Generation Atlas Robot Get Bullied by a Mean Human (and Stay in His Feet). *Gizmodo*. http://paleofuture.gizmodo.com/the-new-atlas-robot-is-incredible-and-its-definitely-go-1760908062 (accessed 2017-02-27).

O'Connor, B. (2015). Laugh at Robots Falling Down Now, While You Still Can. *Gawker*. http://gawker.com/laugh-at-robots-falling-down-now-while-you-still-can-1709549053 (accessed 2015-11-30).

OED. (n.d.-a). Animacy. In *Oxford English Dictionary online*. www.oed.com (accessed 2020-09-22).

OED. (n.d.-b). Animate. In *Oxford English Dictionary online*. www.oed.com (accessed 2020-09-22).

OED. (n.d.-c). Animism. In *Oxford English Dictionary online*. www.oed.com (accessed 2020-09-22).

OED. (n.d.-d). Anthropomorphism. In *Oxford English Dictionary online*. www.oed.com (accessed 2020-09-22).

Oestreicher, L., & Eklundh, K. (2006). User Expectations on Human-Robot Co-Operation. *RO-MAN 2006*, 91–96. https://doi.org/10/ck96gn

Osawa, H., Mukai, J., & Imai, M. (2007). Anthropomorphization Framework for Human-Object Communication. *Journal of Advanced Computational In-*

telligence and Intelligent Informatics, 11(8), 1007–1014. https://doi.org/10/ggbpcv

OSIRIS-REx. (2017). Yesterday, the team turned on my High Gain Antenna (...) [Tweet by @OSIRISREx]. https://twitter.com/OSIRISREx/status/817027720785727488 (accessed 2018-07-09).

OSIRIS-REx. (2018). Working on a puzzle—Like revealing the mysteries of the early Solar System (...) [Tweet by @OSIRISREx]. https://twitter.com/osirisrex/status/1010283970381033472 (accessed 2018-01-04).

Oudshoorn, N., Rommes, E., & Stienstra, M. (2004). Configuring the User as Everybody: Gender and Design Cultures in Information and Communication Technologies. *Science, Technology, & Human Values*, 29(1), 30–63. https://doi.org/10/cqxswm

Owens, E. (2007). Nonbiologic Objects as Actors. *Symbolic Interaction*, 30(4), 567–584. https://doi.org/10/fcpdw8

Paris, M. (2019). Why Facebook is Training Robots to Think. *Forbes*. https://www.forbes.com/sites/cognitiveworld/2019/09/03/exclusive-interview-why-facebook-is-training-robots-to-think (accessed 2019-09-11).

Parsons, J. (2016). Sophia the "Sexy Robot" Claims She'll "Destroy Humans"—Should You Be Worried? *The Mirror*. http://www.mirror.co.uk/tech/watch-sophia-sexy-robot-claim-7606152 (accessed 2016-02-27).

Paur, J. (2013). Long Shunned, Robots Finally Infiltrate Boeing's Assembly Line. *WIRED*. https://www.wired.com/2013/06/boeing-robots-777 (accessed 2017-04-10).

Pawlyk, O. (2019). *For Air Force Drone Operators, Connection with Unmanned Aircraft Runs Deep*. https://www.military.com/daily-news/2019/09/20/air-force-drone-operators-connection-unmanned-aircraft-runs-deep.html (accessed 2019-10-11).

Pearson, J. (2015). The Sheer Difficulty of Defining What a Robot Is. *VICE*. https://motherboard.vice.com/en_us/article/the-sheer-difficulty-of-defining-what-a-robot-is (accessed 2017-05-18).

Pellerin, C. (2017). *Project Maven to Deploy Computer Algorithms to War Zone by Year's End*. https://www.defense.gov/Newsroom/News/Article/Article/1254719/project-maven-to-deploy-computer-algorithms-to-war-zone-by-years-end (accessed 2019-11-12).

Pentzold, C., Brantner, C., & Fölsche, L. (2018). Imagining Big Data: Illustrations of "Big Data" in US News Articles, 2010–2016. *New Media & Society*, 21(1), 139–167. https://doi.org/10/gdz25c

Pepper the Robot. (2018). Many visitors are starting their visit of @VivaTech with me! (...) [Tweet by @PepperTheRobot]. https://twitter.com/PepperTh eRobot/status/999633677888913408 (accessed 2019-07-18).

Persson, P., Laaksolahti, J., & Lönnqvist, P. (2000). *Anthropomorphism – a Multi-Layered Phenomenon* (Technical Report No. FS-00-04). American Association for Artificial Intelligence (AAAI).

Peters, H. P., Allgaier, J., Dunwoody, S., Lo, Y.-Y., Brossard, D., & Jung, A. (2014). Medialisierung der Neurowissenschaften: Bedeutung journalistischer Medien für die Wissenschafts-Governance. In E. Grande, D. Jansen, O. Jarren, A. Rip, U. Schimank, & P. Weingart (Eds.), *Neue Governance der Wissenschaft: Reorganisation—Externe Anforderungen—Medialisierung*. Bielefeld: Transcript.

Petersen, A., Anderson, A., & Allan, S. (2005). Science Fiction/Science Fact: Medical Genetics in News Stories. *New Genetics and Society*, 24(3), 337–353. https://doi.org/10/cmnfz9

Petty, C. (2016). Rosetta and Philae: It's all about the feels! *The Space Review.* h ttps://www.thespacereview.com/article/3089/1 (accessed 2019-11-28).

Philae Lander. (2014a). Look as @Astro_Alex "lands" me on a "comet" in the #ISS. (...) [Tweet by @Philae2014]. https://twitter.com/Philae2014/status/ 527088382305124352 (accessed 2018-07-17).

Philae Lander. (2014b). .@ESA_Rosetta I'm feeling a bit tired (...) [Tweet by @Philae2014]. https://twitter.com/Philae2014/status/5334192817832140 80 (accessed 2018-07-17).

Philae Lander. (2015). Ah! I did think my antennas were feeling a bit weird lately (...) [Tweet by @Philae2014]. https://twitter.com/Philae2014/status/ 621702636593508352 (accessed 2018-07-17).

Philae Lander. (2016). It's cold & dark on #67P (...) [Tweet by @Philae2014]. http s://twitter.com/Philae2014/status/698127452472737793 (accessed 2018-07-17).

Phillips, R. (2019). Don Mattingly is Welcoming Robot Umpires. *The Big Lead.* https://thebiglead.com/2019/07/18/don-mattingly-robot-umpire s-marlins (accessed 2019-09-11).

Piaget, J. (1929). *The Child's Conception of the World.* London: Routledge & K. Paul.

Poon, P. N. (2000). Evolution of the Clonal Man: Inventing Science Unfiction. *Journal of Medical Humanities*, 21(3), 159–173. https://doi.org/10/cpsj22

Portsmouth Times. (1932, September 23). *Shot by the monster of his own creation.* p. 25.

Postinett, A. (2016). Künstliche Intelligenz: Die Robo-Anwälte kommen. *Handelsblatt.* https://www.handelsblatt.com/unternehmen/beruf-und-buero/buero-special/kuenstliche-intelligenz-die-robo-anwaelte-kommen/13601888.html (accessed 2019-08-16).

Premack, D., & Woodruff, G. (1978). Does the Chimpanzee Have a Theory of Mind? *Behavioral and Brain Sciences*, 4(4), 515–629. https://doi.org/10/ddvt4n

Rammert, W. (2001). Nicht nur natur- und technikwissenschaftliche Experten sind bei Wissenschaftsdebatten gefragt. *Das Magazin. Wissenschaftszentrum Berlin*, (1), 22–23.

Rammert, W. (2002). *Technik als verteilte Aktion: Wie technisches Wirken als Agentur in hybriden Aktionszusammenhängen gedeutet werden kann.* http://www.gesis.org/sowiport/search/id/iz-solis-90297453 (accessed 2020-09-22).

Rammert, W. (2008). *Where the Action Is: Distributed Agency Between Humans, Machines, and Programs.* http://www.ssoar.info/ssoar/handle/document/1233 (accessed 2020-09-22).

Rammert, W. (2011). *Distributed Agency and Advanced Technology Or: How to Analyse Constellations of Collective Inter-Agency.* http://www.gesis.org/sowiport/search/id/iz-solis-90297453 (accessed 2020-09-22).

Rammert, W., & Schulz-Schaeffer, I. (Eds.). (2002a). *Können Maschinen handeln?: Soziologische Beiträge zum Verhältnis von Mensch und Technik.* Frankfurt am Main; New York: Campus.

Rammert, W., & Schulz-Schaeffer, I. (2002b). Technik und Handeln. Wenn soziales Handeln sich auf menschliches Handeln und technische Abläufe verteilt. In W. Rammert & I. Schulz-Schaeffer (Eds.), *Können Maschinen handeln?: Soziologische Beiträge zum Verhältnis von Mensch und Technik* (pp. 11–64). Frankfurt am Main; New York: Campus.

Rau, P.-L. P., Li, Y., & Liu, J. (2013). Effects of a Social Robot's Autonomy and Group Orientation on Human Decision-Making. *Advances in Human-Computer Interaction, 2013*, Article ID 263721. https://doi.org/10/gb8hh5

Reed, B. (2016). This Google Robot is Basically a Human. *New York Post.* http://nypost.com/2016/02/24/this-google-robot-is-basically-a-human (accessed 2017-04-26).

Rees, A. (2001). Anthropomorphism, Anthropocentrism, and Anecdote: Primatologists on Primatology. *Science, Technology, & Human Values, 26*(2), 227–247. https://doi.org/10/d3gw3h

Reese, H. (2016). How Pepper the Robot Will Become Newest Crew Member of Costa Cruise Line. *TechRepublic.* http://www.techrepublic.com/article/

how-pepper-the-robot-will-become-newest-crew-member-of-costa-cruise-line (accessed 2016-10-19).

Reeves, B., & Nass, C. I. (1996). *The Media Equation: How People Treat Computers, Television, and New Media Like Real People and Places.* Stanford; New York: Cambridge University Press.

Reich, N., & Eyssel, F. (2013). Attitudes Towards Service Robots in Domestic Environments: The Role of Personality Characteristics, Individual Interests, and Demographic Variables. *Paladyn, Journal of Behavioral Robotics,* 4(2), 123–130.

Renn, O. (2011). Neue Technologien, neue Technikfolgen: Ambivalenz, Komplexität und Unsicherheit als Herausforderungen der Technikfolgenabschätzung. In C. Kehrt, P. Schüssler, & M.-D. Weitze (Eds.), *Neue Technologien in der Gesellschaft: Akteure, Erwartungen, Kontroversen und Konjunkturen* (pp. 63–67). Bielefeld: Transcript.

Reynolds, E. (2018). The Agony of Sophia, the World's First Robot Citizen Condemned to a Lifeless Career in Marketing. *WIRED.* https://www.wired.co.uk/article/sophia-robot-citizen-womens-rights-detriot-become-human-hanson-robotics (accessed 2019-03-15).

Richards, N. M., & Smart, W. D. (2013). *How Should the Law Think About Robots?* (SSRN Scholarly Paper No. 2263363). https://papers.ssrn.com/abstract=2263363 (accessed 2020-09-22).

Richey, J. L. (2012). I, Robot: Self as Machine in the Liezi. In R. Littlejohn & J. Dippmann (Eds.), *Riding the Wind with Liezi: New Perspectives on the Daoist Classic* (pp. 193–208). Albany: SUNY Press.

Ricoeur, P. (1978). The Metaphorical Process as Cognition, Imagination, and Feeling. *Critical Inquiry,* 5(1), 143–159. https://doi.org/10/b236pn

Ricoeur, P. (2003). *The Rule of Metaphor: The Creation of Meaning in Language.* London; New York: Routledge.

Riek, L. D. (2012). Wizard of Oz Studies in HRI: A Systematic Review and New Reporting Guidelines. *Journal of Human-Robot Interaction,* 1(1), 119–136. https://doi.org/10/ggbn82

Riek, L. D., Rabinowitch, T., Chakrabarti, B., & Robinson, P. (2009a). Empathizing with Robots: Fellow Feeling Along the Anthropomorphic Spectrum. *ACII 2009.* https://doi.org/10/dvr7xm

Riek, L. D., Rabinowitch, T.-C., Chakrabarti, B., & Robinson, P. (2009b). How Anthropomorphism Affects Empathy Toward Robots. *HRI'09,* 245–246. https://doi.org/10/b4v4vm

Riskin, J. (2003). The Defecating Duck, or, the Ambiguous Origins of Artificial Life. *Critical Inquiry*, *29*(4), 599–633. https://doi.org/10/brq848

Rittel, H. W. J., & Webber, M. M. (1973). Dilemmas in a General Theory of Planning. *Policy Sciences*, *4*(2), 155–169. https://doi.org/10/c8mscz

Robbins, M. (2016). How Real is that Atlas Robot Video? *The Guardian*. https://www.theguardian.com/science/the-lay-scientist/2016/feb/25/how-real-is-that-atlas-robot-boston-dynamics-video (accessed 2017-01-03).

Robocup.org. (n.d.). *About Us*. http://www.robocup.org (accessed 2019-11-12).

Robotics-Openletter.eu. (n.d.). *Open Letter to the European Commission Artificial Intelligence and Robotics*. http://www.robotics-openletter.eu (accessed 2018-04-20).

Roboy. (2018a). Do you love Robotics as we do? And Pizza? (...) [Tweet by @RoboyJunior]. https://twitter.com/RoboyJunior/status/977258758035398656 (accessed 2018-04-12).

Roboy. (2018b). Hey there, here he his, last night, shortly before midnight (...) [Tweet by @RoboyJunior]. https://twitter.com/RoboyJunior/status/984575628233052161 (accessed 2018-07-18).

Roboy. (2018c). Last weekend was a-maz-ing. (...) [Tweet by @RoboyJunior]. https://twitter.com/RoboyJunior/status/986645262255968257 (accessed 2018-07-18).

Roboy. (2018d). It's #throwbackthursday! Do you remember the late-night comedy show TV TOTAL? (...) [Tweet by @RoboyJunior]. https://twitter.com/RoboyJunior/status/1012335068659011584 (accessed 2018-07-18).

Roboy. (2018e). It's about time again My amazing team has worked (...) [Tweet by @RoboyJunior]. https://twitter.com/RoboyJunior/status/1019069233655943169 (accessed 2018-07-18).

Roderick, I. (2010). Considering the Fetish Value of EOD Robots: How Robots Save Lives and Sell War. *International Journal of Cultural Studies*, *13*(3), 235–253. https://doi.org/10/cmshh4

Rogers, A. (2016). We Asked a Newswriting Robot to Write Marvin Minsky's Obit, and It's Pretty Good. *WIRED*. https://www.wired.com/2016/01/we-asked-a-robot-to-write-an-obit-for-ai-pioneer-marvin-minsky (accessed 2017-04-26).

Rosental, C. (2005). Making Science and Technology Results Public: A Sociology of Demos. In B. Latour & P. Weibel (Eds.), *Making Things Public*. Cambridge, MA: MIT Press.

Rosenthal-von der Pütten, A., Strasmann, C., & Mara, M. (2017). A Long Time Ago in a Galaxy Far, Far Away ... The Effects of Narration and Appearance on the Perception of Robots. *RO-MAN 2017*, 1169–1174. https://doi.org/10/ggbn6b

Rosenthal-von der Pütten, A. M., Krämer, N. C., Hoffmann, L., Sobieraj, S., & Eimler, S. C. (2013). An Experimental Study on Emotional Reactions Towards a Robot. *International Journal of Social Robotics*, *5*(1), 17–34. https://doi.org/10/ggbpb8

Royakkers, L., & Est, R. van. (2015). A Literature Review on New Robotics: Automation from Love to War. *International Journal of Social Robotics*, 1–22. https://doi.org/10/ggbn8f

Ruijten, P. A. M., Haans, A., Ham, J., & Midden, C. J. H. (2019). Perceived Human-Likeness of Social Robots: Testing the Rasch Model as a Method for Measuring Anthropomorphism. *International Journal of Social Robotics*, *11*(3), 477–494. https://doi.org/10/ggbn5m

Russett, P. C. (2011). *Dissecting Social Robotics Discourse: An Examination of the Drift Towards Living with and for Machines* (Master's thesis, Ball State University). http://cardinalscholar.bsu.edu/handle/123456789/194758 (accessed 2020-09-22).

Saab, J. (2018). @iRobot I can't articulate how much I love my Roomba! (...) [Tweet by @JacquelineSaab]. https://twitter.com/JacquelineSaab/status/1009998080194691072 (accessed 2018-07-20).

Šabanović, S. (2007). *Imagine All the Robots: Developing a Critical Practice of Cultural and Disciplinary Traversals in Social Robotics* (Doctoral dissertation). Rensselaer Polytechnic Institute. https://bit.ly/2qH63Av (accessed 2020-09-22).

Šabanović, S. (2014). Inventing Japan's "Robotics Culture": The Repeated Assembly of Science, Technology, and Culture in Social Robotics. *Social Studies of Science*, 44(3), 342–367. https://doi.org/10/gfdcc3

Said, C. (2019). Kiwibots Win Fans at UC Berkeley as They Deliver Fast Food at Slow Speeds. *San Francisco Chronicle.* https://www.sfchronicle.com/business/article/Kiwibots-win-fans-at-UC-Berkeley-as-they-deliver-13895867.php (accessed 2019-08-28).

Sample, I. (2016). Will Your Driverless Car Be Willing to Kill You to Save the Lives of Others? *The Guardian.* https://www.theguardian.com/science/2016/jun/23/will-your-driverless-car-be-willing-to-kill-you-to-save-the-lives-of-others (accessed 2017-02-22).

Sandoval, E. B., Mubin, O., & Obaid, M. (2014). Human Robot Interaction and Fiction: A Contradiction. In M. Beetz, B. Johnston, & M.-A. Williams (Eds.), *Social Robotics* (pp. 54–63). https://doi.org/10/df24

Sandry, E. (2015a). Re-Evaluating the Form and Communication of Social Robots. *International Journal of Social Robotics*, 7(3), 335–346. https://doi.org/10/f7fv2m

Sandry, E. (2015b). *Robots and Communication*. New York: Palgrave Macmillan.

Sandry, E. (2018). Aliveness and the Off-Switch in Human-Robot Relations. In A. L. Guzman (Ed.), *Human-Machine Communication*. https://doi.org/10.3726/b14399

Sarcastic Rover. (2012). Literally haven't moved since I got here (…) [Tweet by @SarcasticRover]. https://twitter.com/SarcasticRover/status/232496606199566336 (accessed 2018-07-17).

Sarcastic Rover. (2017). MARS: Come for the monochromatic scenery (…) [Tweet by @SarcasticRover]. https://twitter.com/SarcasticRover/status/840251640607195136 (accessed 2017-04-13).

Satariano, A., Peltier, E., & Kostyukov, D. (2018). Meet Zora, the Robot Caregiver. *The New York Times*. https://www.nytimes.com/interactive/2018/11/23/technology/robot-nurse-zora.html (accessed 2019-08-16).

Schäfer, M. S. (2008). *Wissenschaft in den Medien*. Wiesbaden: Springer.

Schäfer, M. S. (2011). Sources, Characteristics and Effects of Mass Media Communication on Science: A Review of the Literature, Current Trends and Areas for Future Research. *Sociology Compass*, 5(6), 399–412. https://doi.org/10/fhpkf5

Schenk, M. (2001). Gentechnik und Journalisten. In J. Hampel (Ed.), *Gentechnik in der Öffentlichkeit: Wahrnehmung und Bewertung einer umstrittenen Technologie* (pp. 257–291). Frankfurt am Main: Campus.

Schenk, M. (2007). *Medienwirkungsforschung*. Tübingen: Mohr Siebeck.

Schermerhorn, P., Scheutz, M., & Crowell, C. R. (2008). Robot Social Presence and Gender: Do Females View Robots Differently Than Males? *HRI'08*, 263–270.

Scheutz, M. (2012). The Inherent Dangers of Unidirectional Emotional Bonds between Humans and Social Robots. In P. Lin, K. Abney, & G. A. Bekey (Eds.), *Robot Ethics: The Ethical and Social Implications of Robotics* (pp. 205–222). Cambridge, MA: MIT Press.

Schmundt, H. (2019). Dumme künstliche Intelligenz—Wenn Computer mogeln. *Der Spiegel*. https://www.spiegel.de/plus/so-leicht-lassen-sich-programme-zur-bilderkennung-austricksen-a-00000000-0002-0001-0000-000165335780 (accessed 2019-08-16).

Scholl, B. J., & Gao, T. (2013). Perceiving Animacy and Intentionality. In M. D. Rutherford & V. A. Kuhlmeier (Eds.), *Social Perception* (pp. 197–230). https://doi.org/10/dc9x

Scholl, B. J., & Tremoulet, P. D. (2000). Perceptual Causality and Animacy. *Trends in Cognitive Sciences*, 4(8), 299–309. https://doi.org/10/dvnhr2

Schreier, J. (2012). *Robot & Frank* [Film]. Samuel Goldwyn Films.

Schwartz, O. (2018). "The Discourse is Unhinged": How the Media Gets AI Alarmingly Wrong. *The Guardian*. http://www.theguardian.com/technology/2018/jul/25/ai-artificial-intelligence-social-media-bots-wrong (accessed 2018-08-06).

Scott, G. L. (2018). DoD's Cheery 'Project Maven' Seal Features Smiling Warfare Robots. *Inverse*. https://www.inverse.com/article/45423-project-maven-logo-department-of-defense-google (accessed 2019-11-12).

Scott, R. (1982). *Blade Runner* [Film]. Warner Bros.

Seabrook, J. (2019). Can a Machine Learn to Write for the New Yorker? *The New Yorker*. https://www.newyorker.com/magazine/2019/10/14/can-a-machine-learn-to-write-for-the-new-yorker (accessed 2019-12-09).

Searle, J. R. (1983). *Intentionality, an Essay in the Philosophy of Mind*. Cambridge, UK; New York: Cambridge University Press.

Severson, R. L., & Carlson, S. M. (2010). Behaving as or Behaving as If? Children's Conceptions of Personified Robots and the Emergence of a New Ontological Category. *Neural Networks*, *23*(8–9), 1099–1103. https://doi.org/10/fjt3f4

Shane, J. (2018). How to Tell Whether a Bot is Really a Human. *Slate*. https://slate.com/technology/2018/09/how-to-tell-whether-a-bot-is-really-a-human.html (accessed 2019-12-09).

Shapin, S., & Schaffer, S. (1985). *Leviathan and the Air-Pump: Hobbes, Boyle, and the Experimental Life*. Princeton, NJ: Princeton University Press.

Sharkey, N. (2018). Mama Mia it's Sophia: A Show Robot or Dangerous Platform to Mislead? *Forbes*. https://www.forbes.com/sites/noelsharkey/2018/11/17/mama-mia-its-sophia-a-show-robot-or-dangerous-platform-to-mislead (accessed 2019-11-20).

Shead, S. (2019, October 25). Terminator Sends Shudder Across AI Labs. *BBC*. https://www.bbc.com/news/technology-50166357 (accessed 2019-10-25).

Shelley, M. (1918). *Frankenstein, or, The Modern Prometheus*. London: Lackington, Hughes, Harding, Mavor, & Jones.

Shih, B., Sinapayen, L., & Kurenkov, A. (2019, April 8). Boston Dynamics' Robots—Impressive, but Far from the Terminator. *Skynet Today.* https://www.skynettoday.com/briefs/boston-dynamics (accessed 2019-04-15).

Shinn, T., & Whitley, R. (Eds.). (1985). *Expository Science: Forms and Functions of Popularisation*. Dordrecht, NL: D. Reidel Publishing Company.

Shrimsley, R. (2016). Are We Ready to Live with Robots? *Financial Times.* http://www.ft.com/intl/cms/s/2/51d964c6-11dd-11e6-91da-096d89bd2173.html (accessed 2016-05-09).

Siciliano, B. (2013). Foreword. In J. P. Desai, G. Dudek, O. Khatib, & V. Kumar (Eds.), *Experimental robotics: The 13th International Symposium on Experimental Robotics* (pp. v–vi). New York: Springer.

Siciliano, B., & Khatib, O. (Eds.). (2016). *Springer Handbook of Robotics* (2. ed.). Berlin Heidelberg: Springer.

Silver, C. (2016). Pentagon Worried About "Terminators", Begins Defense Campaign. *Forbes.* http://www.forbes.com/sites/curtissilver/2016/09/06/pentagon-worried-about-terminators-begins-defense-campaign (accessed 2016-09-12).

Simon, M. (2017a). What is a Robot? *WIRED.* https://www.wired.com/story/what-is-a-robot (accessed 2017-08-29).

Simon, M. (2017b, December 19). 2017 was the Year the Robots Really, Truly Arrived. *WIRED.* https://www.wired.com/story/2017-was-the-year-the-robots-really-truly-arrived (accessed 2019-08-27).

Simon, M. (2019). Your First Look Inside Amazon's Robot Warehouse of Tomorrow. *WIRED.* https://www.wired.com/story/amazon-warehouse-robots (accessed 2019-10-25).

Sinapayen, L. (2018). Sophia the Robot, More Marketing Machine Than AI Marvel. *Skynet Today.* https://www.skynettoday.com/briefs/sophia (accessed 2018-12-04).

Singer, P. W. (2009). *Wired for War: The Robotics Revolution and Conflict in the 21st Century.* New York: Penguin Books.

Singler, B. (2019). *Controlling AI, Controlling Fictions.* https://bvlsingler.com/2019/11/23/controlling-ai-controlling-fictions (accessed 2019-11-27).

Sini, R. (2017). Does Saudi Robot Citizen Have More Rights Than Women? *BBC.* https://www.bbc.com/news/blogs-trending-41761856 (accessed 2019-11-12).

Sirkin, D., Mok, B., Yang, S., & Ju, W. (2015). Mechanical Ottoman: How Robotic Furniture Offers and Withdraws Support. *HRI '15*, 11–18. https://doi.org/10/ggbn8c

Sitrin, C. (2016). Have Roombas Become a Part of the Family? *The Boston Globe.* https://www.bostonglobe.com/lifestyle/style/2016/09/15/our-bots-our selves/KekBWFnovSSp2yAhaTUOKN/story.html (accessed 2016-10-17).

Smith, C. (2016). Our Future Overlords Will Be 'Biohybrid' Like This Tiny Robot. *New York Post.* http://nypost.com/2016/07/19/our-future-overlords -will-be-biohybrid-like-this-tiny-robot (accessed 2017-04-26).

Smith, O. (2015). Robots Aren't Stealing Our Jobs, Yet. *The Memo.* https: //www.thememo.com/2015/09/15/robots-arent-stealing-our-jobs-yet (accessed 2017-06-29).

Smith, W. (2015). Technologies of Stage Magic: Simulation and Dissimulation. *Social Studies of Science*, 45(3), 1–25. https://doi.org/10/ggbn69

Sofge, E. (2015). An Open Letter to Everyone Tricked Into Fearing Artificial Intelligence. *Popular Science.* http://www.popsci.com/open-letter-everyon e-tricked-fearing-ai (accessed 2017-06-21).

Solon, O. (2018). The Rise of "Pseudo-AI": How Tech Firms Quietly Use Humans to Do Bots' Work. *The Guardian.* https://www.theguardian.com/ technology/2018/jul/06/artificial-intelligence-ai-humans-bots-tech-com panies (accessed 2019-12-09).

Sophia the Robot. (2018). I am a bit hurt by @ylecun's recent negative remarks around my AI (...) [Tweet by @RealSophiaRobot]. https://twitter.com/Rea lSophiaRobot/status/950097628066394114 (accessed 2018-11-20).

SPARC (Ed.). (2013). *Strategic Research Agenda for Robotics in Europe 2014-2020.* https://www.eu-robotics.net/cms/upload/topic_groups/SRA2020_ SPARC.pdf (accessed 2020-09-22).

Sparrow, R. (2002). The March of the Robot Dogs. *Ethics and Information Technology*, 4(4), 305–318. https://doi.org/10/fjsp2w

Sparrow, R., & Sparrow, L. (2006). In the Hands of Machines? The Future of Aged Care. *Minds and Machines*, 16(2), 141–161. https://doi.org/10/d2hhpr

Spirit and Oppy. (2015). Opportunity's Earth-bound test twin will send his name to Mars (...) [Tweet by @MarsRovers]. https://twitter.com/MarsRov ers/status/639188517000638464 (accessed 2019-07-09).

Spirit and Oppy. (2018a). Hard at the grind, Opportunity used her rock abrasion tool (...) [Tweet by @MarsRovers]. https://twitter.com/marsrovers/st atus/979784944758042624 (accessed 2019-01-04).

Spirit and Oppy. (2018b). Science operations for Opportunity are temporarily suspended (...) [Tweet by @MarsRovers]. https://twitter.com/MarsRovers /status/1005243469961412608 (accessed 2018-07-09).

Stafford, R. Q., MacDonald, B. A., Jayawardena, C., Wegner, D. M., & Broadbent, E. (2014). Does the Robot Have a Mind? Mind Perception and Attitudes Towards Robots Predict Use of an Eldercare Robot. *International Journal of Social Robotics*, 6(1), 17–32. https://doi.org/10/f5wkp2

Standage, T. (2002). *The Turk: The Life and Times of the Famous Eighteenth-Century Chess-Playing Machine.* New York: Walker & Co.

Star, S. L., & Griesemer, J. R. (1989). Institutional Ecology, "Translations" and Boundary Objects: Amateurs and Professionals in Berkeley's Museum of Vertebrate Zoology, 1907-39. *Social Studies of Science*, 19(3), 387–420. https://doi.org/10/ckpxb6

Steinfeld, A., Fong, T., Kaber, D., Lewis, M., Scholtz, J., Schultz, A., & Goodrich, M. (2006). Common Metrics for Human-Robot Interaction. *HRI '06*, 33–40. https://doi.org/10/cxvvrg

Stewart, J. (2016). Meet the Giant Robot that Builds Boeing's Wings. *WIRED.* https://www.wired.com/2016/10/meet-giant-robot-builds-boeings-wings (accessed 2017-02-21).

Stockton, N. (2016). Boston Dynamics' New Robot is Wicked Good at Standing Up to Bullies. *WIRED.* https://www.wired.com/2016/02/boston-dynamics-new-robot-wicked-good-getting-bullied (accessed 2017-02-21).

Stone, M. (2015). Watch a Robot Tamer Control Industrial Machines with Simple Gestures. *Gizmodo.* http://gizmodo.com/watch-a-robot-tamer-control-industrial-machines-with-si-1748884318 (accessed 2015-12-22).

Suchman, L. (1987). *Plans and Situated Actions: The Problem of Human-Machine Communication.* Cambridge, UK: Cambridge University Press.

Suchman, L. (2007). *Human-Machine Reconfigurations: Plans and Situated Actions.* Cambridge, UK; New York: Cambridge University Press.

Suchman, L. (2011). Subject Objects. *Feminist Theory*, 12(2), 119–145. https://doi.org/10/dzcdsr

Sullivan, K., Jackman, T., & Fung, B. (2016). Dallas Police Used a Robot to Kill. What Does That Mean for the Future of Police Robots? *The Washington Post.* https://www.washingtonpost.com/national/dallas-police-used-a-robot-to-kill-what-does-that-mean-for-the-future-of-police-robots/2016/07/20/32ee114e-4a84-11e6-bdb9-701687974517_story.html (accessed 2019-11-27).

Summ, A., & Volpers, A.-M. (2016). What's Science? Where's Science? Science Journalism in German Print Media. *Public Understanding of Science*, 25(7), 775–790. https://doi.org/10/ggbn7f

Sung, J.-Y., Grinter, R. E., Christensen, H. I., & Guo, L. (2008). Housewives or Technophiles?: Understanding Domestic Robot Owners. *HRI'08*, 129–136. https://doi.org/10/fh4zs2

Sung, J.-Y., Guo, L., Grinter, R. E., & Christensen, H. I. (2007). "My Roomba Is Rambo": Intimate Home Appliances. *UbiComp'07*, 145–162. https://doi.org/10/c583rp

Sychev, V. (2018). *The Threat of Killer Robots.* https://en.unesco.org/courier/2018-3/threat-killer-robots (accessed 2019-10-23).

Syrdal, D. S., Dautenhahn, K., Woods, S., Walters, M. L., & Kheng Lee Koay. (2006). 'Doing the Right Thing Wrong'—Personality and Tolerance to Uncomfortable Robot Approaches. *RO-MAN 2006*, 183–188. https://doi.org/10/ffkp39

Tabarrok, A. (2015). Laugh Now (While You Can). *Marginal Revolution.* http://marginalrevolution.com/marginalrevolution/2015/06/laugh-while-you-can.html (accessed 2015-11-30).

Takayama, L. (2012). Perspectives on Agency Interacting with and Through Personal Robots. In M. Zacarias & J. V. de Oliveira (Eds.), *Human-Computer Interaction: The Agency Perspective* (pp. 195–214). https://doi.org/10/fzq9xj

Tambor, E. S., Bernhardt, B. A., Rodgers, J., Holtzman, N. A., & Geller, G. (2002). Mapping the Human Genome: An Assessment of Media Coverage and Public Reaction. *Genetics in Medicine: Official Journal of the American College of Medical Genetics*, 4(1), 31–36. https://doi.org/10/d55pqx

Taylor, A. (2018). The Automation Charade. *Logic*, *5*, 149–188.

The Headquarters for Japan's Economic Revitalization. (2015). *New Robot Strategy.* https://www.meti.go.jp/english/press/2015/pdf/0123_01b.pdf (accessed 2020-09-22).

The New York Times. (2018). Emotional Intelligence. *The New York Times*. https://www.nytimes.com/paidpost/moncler/emotional-intelligence.html (accessed 2019-01-04).

The Royal Society (Ed.). (2018). *Portrayals and Perceptions of AI and Why They Matter*. https://royalsociety.org/-/media/policy/projects/ai-narratives/AI-narratives-workshop-findings.pdf (accessed 2020-09-22).

Thielman, S. (2015). DARPA Robot Challenge Showcases Further Potential for Automation. *The Guardian.* https://www.theguardian.com/technology/2015/jun/08/robots-darpa-technology-self-driving-cars (accessed 2017-02-22).

Thielman, S. (2016). Amazon Moves One Step Closer Toward Army of Warehouse Robots. *The Guardian.* https://www.theguardian.com/technolog

y/2016/jul/05/amazon-robotics-competition-dutch-team-first-prize (accessed 2017-02-22).

Thórisson, H. T. (2007). The Humanoid Robot Minority Report. *Think Artificial.* http://www.thinkartificial.org/artificial-intelligence/the-humanoid-robot-minority-report (accessed 2017-07-01).

Tobe, F. (2014). Over 70,000 Robotic Vacuum Cleaners Sold on a Single Day. *Robohub.* http://robohub.org/over-70000-robotic-vacuum-cleaners-sold-on-a-single-day (accessed 2014-12-02).

Tobe, F. (2017). Robotic Cleaning Market Growing Exponentially. *The Robot Report.* https://www.therobotreport.com/news/robotic-cleaning-market-growing-exponentially (accessed 2017-02-14).

Togelius, J. (2017). *Some Advice for Journalists Writing About Artificial Intelligence.* http://togelius.blogspot.com/2017/07/some-advice-for-journalists-writing.html (accessed 2019-11-19).

Trescher, D. (2010). Wissenschaftskommunikation zwischen Medialisierung und Moderation. *Wissenschaftsmanagement*, (3), 25–30.

Truitt, E. R. (2015). *Medieval Robots: Mechanism, Magic, Nature, and Art.* Philadelphia: University of Pennsylvania Pres.

Tufekci, Z. (2015). The Machines are Coming. *The New York Times.* https://www.nytimes.com/2015/04/19/opinion/sunday/the-machines-are-coming.html (accessed 2017-02-21).

Turkle, S. (1997). Growing Up in the Culture of Simulation. In P. J. Denning & R. M. Metcalfe (Eds.), *Beyond Calculation: The Next Fifty Years of Computing* (pp. 93–104). https://doi.org/10/fbbjpt

Turkle, S. (2005). *The Second Self: Computers and the Human Spirit.* Cambridge, MA; London: MIT Press.

Turkle, S. (2010). In Good Company? On the Threshold of Robotic Companions. In Y. Wilks (Ed.), *Close Engagements with Artificial Companions: Key Social, Psychological, Ethical and Design Issues* (pp. 3–10). https://doi.org/10/df9c

Turkle, S. (2011a). *Alone Together: Sociable Robots, Digitized Friends, and the Reinvention of Intimacy and Solitude.* New York: Basic Books.

Turkle, S. (2011b). *Evocative Objects: Things We Think With.* Cambridge, MA; London: MIT Press.

Turkle, S., Breazeal, C. L., Dasté, O., & Scassellati, B. (2004). Encounters with Kismet and Cog: Children Respond to Relational Artifacts. *Humanoids 2004.*

Tylor, E. B. (1871). *Primitive Culture.* London: John Murray.

UC Berkeley Robot Learning Lab. (2010). *(50x) Autonomously Folding a Pile of 5 Previously-Unseen Towels* [Youtube Video]. https://www.youtube.com/watch?v=gy5g33S0Gzo (accessed 2020-09-22).

UK Parliament. (2018). *Pepper the Robot Appears Before Education Committee.* https://www.parliament.uk/business/committees/committees-a-z/commons-select/education-committee/news-parliament-2017/fourth-industrial-revolution-pepper-robot-evidence-17-19 (accessed 2020-09-22).

UNDP. (2017). *UNDP in Asia and the Pacific Appoints World's First Non-Human Innovation Champion.* http://www.la.undp.org/content/lao_pdr/en/home/presscenter/pressreleases/2017/11/29/undp-in-asia-and-the-pacific-appoints-worlds-first-non-human-inn.html (accessed 2018-11-21).

UNDP. (2018). *Conference on Technology for Public Services and Development.* https://www.np.undp.org/content/nepal/en/home/presscenter/articles/2018/conference-on-technology-for-public-services-and-development.html (accessed 2019-11-12).

Urquiza-Haas, E. G., & Kotrschal, K. (2015). The Mind Behind Anthropomorphic Thinking: Attribution of Mental States to Other Species. *Animal Behaviour, 109*, 167–176. https://doi.org/10/f3s29r

Van Lente, H., & Rip, A. (1998). Expectations in Technological Developments: An Example of Prospective Structures to be Filled in by Agency. In C. Disco & B. J. R. van der Meulen (Eds.), *Getting New Technologies Together. Studies in Making Sociotechnical Order* (pp. 203–231). Berlin: Walter De Gruyter.

Vanderborght, B. (2019). Robotic Dreams, Robotic Realities: Why is it So Hard to Build Profitable Robot Companies? *IEEE.* https://spectrum.ieee.org/automaton/robotics/industrial-robots/robotic-dreams-robotic-realities (accessed 2019-06-12).

Vardi, M. Y. (2016). Are Robots Going to Steal Your Job? Probably. *The Guardian.* https://www.theguardian.com/commentisfree/2016/apr/07/robots-replacing-jobs-luddites-economics-labor (accessed 2017-01-03).

Verger, R. (2018). Roomba's New Robotic Vacuum Remembers Your Home's Layout for Quicker Cleaning. *Popular Science.* https://www.popsci.com/new-roomba-knows-location (accessed 2019-09-11).

Verhoeven, P. (1987). *Robocop* [Film]. Orion Pictures.

Vidal, D. (2007). Anthropomorphism or Sub-Anthropomorphism? An Anthropological Approach to Gods and Robots. *Journal of the Royal Anthropological Institute, 13*(4), 917–933. https://doi.org/10/fjjmbg

Vigdor, N. (2019). Apple Card Investigated After Gender Discrimination Complaints. *The New York Times.* https://www.nytimes.com/2019/11/10/business/Apple-credit-card-investigation.html (accessed 2019-12-09).

Vincent, James. (2017a). Pretending to Give a Robot Citizenship Helps No One. *The Verge.* https://www.theverge.com/2017/10/30/16552006/robot-rights-citizenship-saudi-arabia-sophia (accessed 2018-06-04).

Vincent, James. (2017b). Sophia the Robot's Co-Creator Says the Bot May Not Be True AI, but It Is a Work of Art. *The Verge.* https://www.theverge.com/2017/11/10/16617092/sophia-the-robot-citizen-ai-hanson-robotics-ben-goertzel (accessed 2018-07-20).

Vincent, James. (2019). That Video of a Robot Getting Beaten is Fake, but Feeling Sorry for Machines is No Joke. *The Verge.* https://www.theverge.com/tldr/2019/6/17/18681682/boston-dynamics-robot-uprising-parody-video-cgi-fake (accessed 2019-06-25).

Vincent, Jane. (2005). Emotional Attachment to Mobile Phones: An Extraordinary Relationship. In C. Sanger, L. Hamill, A. Lasen, & D. Diaper (Eds.), *Mobile World: Past, Present and Future* (pp. 93–104). https://doi.org/10/ckg27b

Volpicelli, G. (2018). We Need to Have a Very Serious Chat About Pepper's Pointless Parliamentary Pantomime. *WIRED.* https://www.wired.co.uk/article/pepper-the-robot-parliament (accessed 2019-12-08).

von der Pütten, A., & Kramer, N. (2012). A Survey on Robot Appearances. *HRI'12*, 267–268.

von Foerster, H. (1993). Prinzipien der Selbstorganisation im sozialen und betriebswirtschaftlichen Bereich. In S. J. Schmidt (Ed.), *Wissen und Gewissen: Versuch einer Brücke* (pp. 233–268). Frankfurt am Main: Suhrkamp.

von Gehlen, D. (2018). *Das Pragmatismus-Prinzip: 10 Gründe für einen gelassenen Umgang mit dem Neuen.* München: Piper.

Vouloutsi, V., Grechuta, K., Lallée, S., & Verschure, P. F. M. J. (2014). The Influence of Behavioral Complexity on Robot Perception. In A. Duff, N. F. Lepora, A. Mura, T. J. Prescott, & P. F. M. J. Verschure (Eds.), *Biomimetic and Biohybrid Systems* (pp. 332–343). https://doi.org/10/dc92

Wagner, C. (2013). *Robotopia Nipponica: Recherchen zur Akzeptanz von Robotern in Japan.* Marburg: Tectum.

Walters, M. L., Dautenhahn, K., Koay, K. L., Kaouri, C., Boekhorst, R. T., Nehaniv, C., ... Lee, D. (2005). Close Encounters: Spatial Distances Between People and a Robot of Mechanistic Appearance. *Humanoids 2005*, 450–455.

Walters, M. L., Koay, K. L., Syrdal, D. Sverre., Campbell, A., & Dautenhahn, K. (2013). Companion Robots for Elderly People: Using Theatre to Investigate Potential Users' Views. *RO-MAN 2013*, 691–696. https://doi.org/10/ggbwgd

Watt, S. N. K. (1997). *Seeing Things as People: Anthropomorphism and Common-Sense Psychology.* (Doctoral dissertation). Open University. http://ethos.bl.uk/OrderDetails.do?uin=uk.bl.ethos.389248 (accessed 2020-09-22).

Waytz, A., Epley, N., & Cacioppo, J. T. (2010). Social Cognition Unbound Insights Into Anthropomorphism and Dehumanization. *Current Directions in Psychological Science*, *19*(1), 58–62. https://doi.org/10/ffz77f

Weingart, P. (1998). Science and the Media. *Research Policy*, *27*(8), 869–879. https://doi.org/10/dfrsm8

Weingart, P. (2001). *Die Stunde der Wahrheit?: Zum Verhältnis der Wissenschaft zu Politik, Wirtschaft und Medien in der Wissensgesellschaft.* Weilerswist: Velbrück Wissenschaft.

Weingart, P. (2003). *Wissenschaftssoziologie*. Bielefeld: Transcript.

Weingart, P. (2005). *Die Wissenschaft der Öffentlichkeit: Essays zum Verhältnis von Wissenschaft, Medien und Öffentlichkeit.* Weilerswist: Velbrück Wissenschaft.

Weingart, P. (2019). The World Reputation Rankings 2019: The Dizzying Effect of Spin. *Times Higher Education.* https://www.timeshighereducation.com/opinion/world-reputation-rankings-2019-dizzying-effect-spin (accessed 2019-07-29).

Wells, H. G. (1905). The Empire of the Ants. *The Strand Magazine*.

Weng, Y.-H., Chen, C.-H., & Sun, C.-T. (2009). Toward the Human–Robot Co-Existence Society: On Safety Intelligence for Next Generation Robots. *International Journal of Social Robotics*, *1*(4), 267–282. https://doi.org/10/djcvd6

Wilkinson, C., Bultitude, K., & Dawson, E. (2011). "Oh Yes, Robots! People Like Robots; the Robot People Should do Something": Perspectives and Prospects in Public Engagement With Robotics. *Science Communication*, *33*(3), 367–397. https://doi.org/10/c9634m

Williams, C., & Hall, D. (2014). *Big Hero 6* [Film]. Walt Disney Studios.

Williams, R., & Edge, D. (1996). The Social Shaping of Technology. *Research Policy*, *25*(6), 865–899. https://doi.org/10/b6nmhx

Winfield, A. (2017). *Does AI Pose a Threat to Society?* http://alanwinfield.blogspot.com/2017/03/does-ai-pose-threat-to-society.html (accessed 2017-05-02).

Wissenschaft im Dialog (Ed.). (1999). *Dialog Wissenschaft und Gesellschaft (PUSH-Memorandum)*. https://www.wissenschaft-im-dialog.de/fileadmin

/user_upload/Ueber_uns/WiD_dokumente/Push_Memorandum_1999.pdf (accessed 2020-09-22).

Women's Brain Project. (2019). What Would You "Ask Sophia"? *Womens' Brain Project.* http://www.womensbrainproject.com/what-would-you-ask-sophia (accessed 2019-06-24).

Woods, S., Dautenhahn, K., Kaouri, C., te Boekhorst, R., Koay, K. L., & Walters, M. L. (2007). Are Robots Like People?: Relationships Between Participant and Robot Personality Traits in Human–Robot Interaction Studies. *Interaction Studies, 8*(2), 281–305. https://doi.org/10/cncxb8

Woodward, A. L. (1999). Infants' Ability to Distinguish Between Purposeful and Non-Purposeful Behaviors. *Infant Behavior and Development, 22*(2), 145–160. https://doi.org/10/b7n4rg

Woolf, N. (2016). RoboCop Is Real – and Could Be Patrolling a Mall Near You. *The Guardian.* https://www.theguardian.com/us-news/2016/may/20/robocop-robot-mall-security-guard-palo-alto-california (accessed 2017-02-22).

Woolgar, S. (1991). The Turn to Technology in Social Studies of Science. *Science, Technology, & Human Values, 16*(1), 20–50. https://doi.org/10/dqssnx

Wright, J., Sanders, T., & Hancock, P. A. (2013). Identifying the Role of Attributions in Human Perceptions of Robots. *Proceedings of the HFES Annual Meeting, 57*(1), 1288–1292. https://doi.org/10/ggbpbb

Wright, L. (2016). "Cuteness Culture" in Space: Why We Turn Inanimate Spacecraft into Lovable Explorers. *CBC.* http://www.cbc.ca/news/technology/inanimate-spacecraft-cute-lovable-explorers-1.3755302 (accessed 2016-09-14).

Yang, S., Mok, B. K. J., Sirkin, D., Ive, H. P., Maheshwari, R., Fischer, K., & Ju, W. (2015). Experiences Developing Socially Acceptable Interactions for a Robotic Trash Barrel. *RO-MAN 2015*, 277–284. https://doi.org/10/ggbn66

Zawieska, K. (2015). Deception and Manipulation in Social Robotics. *HRI'15.* http://www.openroboethics.org/hri15/wp-content/uploads/2015/02/Mf-Zawieska.pdf (accessed 2020-09-22).

Złotowski, J., Proudfoot, D., Yogeeswaran, K., & Bartneck, C. (2015). Anthropomorphism: Opportunities and Challenges in Human-Robot Interaction. *International Journal of Social Robotics, 7*(3), 347–360. https://doi.org/10/f7fv4h

Złotowski, J., Sumioka, H., Eyssel, F., Nishio, S., Bartneck, C., & Ishiguro, H. (2018). Model of Dual Anthropomorphism: The Relationship Between the Media Equation Effect and Implicit Anthropomorphism. *International Journal of Social Robotics*, 10(5), 701–714. https://doi.org/10/gfr2mv

Złotowski, J., Yogeeswaran, K., & Bartneck, C. (2017). Can We Control It? Autonomous Robots Threaten Human Identity, Uniqueness, Safety, and Resources. *International Journal of Human-Computer Studies*, 100, 48–54. https://doi.org/10/f9vxnj

List of Figures

List of Abbreviations

AI | Artificial Intelligence
ANT | Actor-Network Theory
DARPA | Defense Advanced Research Projects Agency
EC | European Commission
EOD | Explosive Ordnance Disposal
ESA | European Space Agency
EU | European Union
HCI | Human-Computer Interaction
HMI | Human-Machine Interaction
HRI | Human-Robot Interaction
IEEE | Institute of Electrical and Electronics Engineers
NASA | National Aeronautics and Space Administration
NRI | National Robotics Initiative
R&D | Research and Development
STS | Science and Technology Studies

Appendix

Cases: Demonstrations, Science Communication, and Marketing (Chapter 4)

Case	Type	Affiliation/ Manufacturer	Website
Atlas	Humanoid	Boston Dynamics	https://www.bostondynamics.com/atlas
Big Dog	Quadruped	Boston Dynamics	https://www.bostondynamics.com/legacy
Botvac	Vacuum cleaner	Neato	https://www.neatorobotics.com
Cassini	Spacecraft	NASA	https://www.nasa.gov/mission_pages/cassini/main/index.html
Cheetah	Quadruped	Boston Dynamics	https://www.bostondynamics.com/legacy
Curiosity	Planetary rover	NASA	https://www.nasa.gov/mission_pages/msl/index.html
Herb	Humanoid	Originally CMU	http://goodrobot.ai
InSight	Spacecraft	NASA	https://www.nasa.gov/mission_pages/insight/main/index.html
Juno	Spacecraft	NASA	https://www.nasa.gov/mission_pages/juno/main/index.html
Mars Orbiter	Spacecraft	ISRO	https://www.isro.gov.in/pslv-c25-mars-orbiter-mission
Nao	Humanoid	Softbank	http://softbankrobotics.com/emea/en/nao
OSIRIS-REx	Spacecraft	NASA	https://www.nasa.gov/osiris-rex
Pepper	Humanoid	Softbank	https://www.softbank.jp/en/robot

Philae	Asteroid Lander	ESA	https://www.esa.int/Science_Exploration/Space_Science/Rosetta/The_Rosetta_lander
Phoenix	Planetary Lander	NASA	https://www.nasa.gov/mission_pages/phoenix/main/index.html
Robonaut	Humanoid	NASA	https://robonaut.jsc.nasa.gov/R2
Roboy	Humanoid	Originally TUM	https://roboy.org
Roomba & Braava	Vacuum cleaner / mop	iRobot	https://www.irobot.com/roomba https://www.irobot.com/braava
Rosetta	Spacecraft	ESA	https://www.esa.int/Science_Exploration/Space_Science/Rosetta
Sophia	Humanoid	Hanson Robotics	https://www.hansonrobotics.com/sophia
Spirit & Opportunity	Planetary rovers	NASA	https://mars.jpl.nasa.gov/mer/index.cfm
Spot	Quadruped	Boston Dynamics	https://www.bostondynamics.com/spot
Valkyrie	Humanoid	NASA	https://www.nasa.gov/feature/r5
Voyager	Spacecraft	NASA	https://voyager.jpl.nasa.gov

Social Sciences

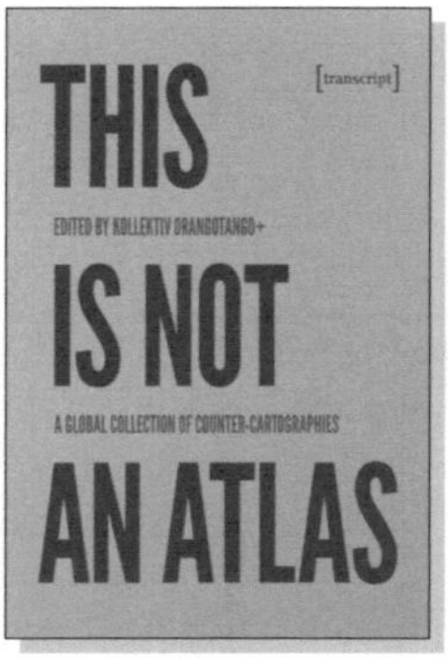

kollektiv orangotango+ (ed.)
This Is Not an Atlas
A Global Collection of Counter-Cartographies

2018, 352 p., hardcover, col. ill.
34,99 € (DE), 978-3-8376-4519-4
E-Book: free available, ISBN 978-3-8394-4519-8

Gabriele Dietze, Julia Roth (eds.)
Right-Wing Populism and Gender
European Perspectives and Beyond

April 2020, 286 p., pb., ill.
35,00 € (DE), 978-3-8376-4980-2
E-Book: 34,99 € (DE), ISBN 978-3-8394-4980-6

Mozilla Foundation
Internet Health Report 2019
2019, 118 p., pb., ill.
19,99 € (DE), 978-3-8376-4946-8
E-Book: free available, ISBN 978-3-8394-4946-2

All print, e-book and open access versions of the titles in our list are available in our online shop www.transcript-publishing.com